INTELLECTUAL

ARITHMETIC,

OR,

AN ANALYSIS OF THE SCIENCE OF NUMBERS

WITH ESPECIAL REFERENCE TO

MENTAL TRAINING AND DEVELOPMENT

BY
CHARLES DAVIES, LL.D.,
AUTHOR OF A SERIES OF ARITHMETICS, ELEMENTARY ALGEBRA,
ELEMENTS OF SURVEYING, ELEMENTS OF DESCRIPTIVE
GEOMETRY, SHADES, SHADOWS, AND PERSPECTIVE,
ANALYTICAL GEOMETRY, AND DIFFERENTIAL
AND INTEGRAL CALCULUS.

NEW YORK:
PUBLISHED BY A. S. BARNES & CO.
BALTIMORE: J. W. BOND & CO.—CINCINNATI: H. W. DERBY.—CHICAGO
D. B. COOK & CO.—ST. LOUIS: L. & A. CARR.—NEW ORLEANS:
W. FLEMMING.—MOBILE: RANDALL & WILLIAMS.
1858.

SUGGESTIONS.

THIS work is designed both for primary and advanced classes. The first part is adapted to beginners, while the latter part is peculiarly fitted to give to the more advanced student that *thorough mental drilling*, in the *Analysis of Numbers*, which furnishes the true basis of all mathematical knowledge.

It is suggested that classes in Higher Arithmetic, and even in Algebra, not familiar with works of this kind, will be greatly benefited by a *thorough* exercise in this most important branch of mathematical science.

The Teacher should require the class to dispense with their books at the time of recitation. He should read each example, and then call upon some member of the class to solve it. The pupil should rise and repeat the example in the same language used by the teacher, and should then proceed to analyze it.

The analysis will be found to consist of three parts; two propositions and a conclusion; thus:

What will 4 barrels of cider cost at 3 dollars a barrel?

1ST PROPOSITION: Four barrels will cost 4 times as much as 1 barrel.

2D PROPOSITION: If 1 barrel costs 3 dollars, 4 barrels will cost 4 times 3 dollars, which are 12 dollars:

CONCLUSION: Therefore, 4 barrels of cider at 3 dollars a barrel, will cost 12 dollars.

The pupil should *never* be allowed to omit either of the steps; and he should be required always to adhere strictly to a *correct* and *uniform phraseology* in the analysis.

The *forms* of analysis are thought to be of great service both to the teacher and pupil.

It is also suggested, that the pupil be thoroughly drilled in Lessons III. and IV., Sect. VII., as they afford very valuable mental exercises and a great variety of Arithmetical processes.

INTRODUCTION.

Every book of instruction should have a specific object to which the entire work, both in matter and method, should strictly conform.

It is the object of this book to train and develop the mind by means of the science of numbers. Numbers are the instruments here employed to strengthen the memory, to cultivate the faculty of abstraction and to give force and vigor to the reasoning powers.

All our ideas of numbers are either of unity or of multiplicity—unity being the elementary idea from which all others are derived.

A true analysis must conform to the nature of the subject analyzed. It must separate all the ideas and principles into their primary elements, and then explain and make manifest the laws by which these elements are connected with each other. Hence, the analysis of numbers must begin with the unit 1,—for this is the foundation, and the science is but the development of the various processes by which all other numbers are derived from 1, as a base, and a comparison of the base 1, with the numbers so derived.

Every number has what we call a *base:* that is, "number being a collection of things of the same kind," one of these things is the base of the number; and *this thing*, is called a *unit.* If we have the number 3 hundred, we may consider it in several points of view:

1*st.* It is one hundred taken 3 times, and if we regard one hundred as the base, then, the base is taken 3 times to make up the number; and 100 is the unit.

2*nd.* We may consider the number as made up of 30 tens, and if we regard 10 as the base, then the base is taken 30 times; and 10 is the unit.

3*rd.* We may also consider the number as made up of 300 ones, in which, the base is 1, and the unit of the number 1.

Again, if we analyze the number,

cwt.	*qr.*	*lb.*	*oz.*	*dr.*
13	2	20	12	4

We see, that 1*cwt.* is the base of 13*cwt.*; 1*qr.* the base of 2*qr.*; 1*lb.* the base of 20*lb.*; 1*oz.* the base of 12*oz.*; and *dr.* the base of 4*dr.*, and all these bases may be referred to 1 dram as a primary base; hence, as in simple numbers, every base may be referred to the unit 1: therefore, *in every entire number*, 1 *is the primary base.*

Let us see if the same be true in fractional numbers.

If we have the fraction $\frac{7}{8}$ it denotes:

1*st.* That something regarded as a whole has been divided into 8 equal parts: and,

2*nd.* That 7 of these parts are taken.

In this collection of 7 things, (each of which is $\frac{1}{8}$), $\frac{1}{8}$ is the base of the fractional number $\frac{7}{8}$; but it is not the primary base; for $\frac{1}{8}$ implies, either $\frac{1}{8}$ of 1 or $\frac{1}{8}$ of some collection of 1's; if a collection of 1's we call that collection *unity*, which may be referred to the primary base 1: hence, *every number, either integral or fractional, has the unit* 1 *for a primary base.*

A fractional number, therefore, is merely a collection of the equal parts of unity, and to one of these parts we give the name of *fractional unit*. The unit which is divided is called the *unit of the fraction*, and may be a collection of units, (as what is $\frac{1}{2}$ of 40?) or it may be the unit 1.

The term UNITY, in mathematical science, is applied to any number or quantity regarded as a whole: the term *unit*, in arithmetic, to any number which is used as the base of a collection. Thus, 10 is a unit of the second order, being the base for the collection of 10's 100 is a unit of the third order, being the base for the collection of hundreds, and similarly for other bases. Thus, also, in the fraction $\frac{7}{8}$, $\frac{1}{8}$ is the fractional unit, being the fractional base, while the primary base is the unit 1.

Every arithmetical process, therefore, has a direct reference to the unit 1; and with this view of the subject before him, the pupil always has the means of making a correct analysis.

Addition is the process of finding a number which shall contain as many units, and no more, as are found in all the numbers added. Multiplication is taking one number, called the multiplicand, as many times as there are units in another number, called the multiplier, and the number which shows the result of such taking, is called the product: and similarly for all other arithmetical processes.

A clear conception of elementary principles, by which we mean, those principles that result from a final analysis, lies at the foundation of all knowledge. It is not till we get such conceptions, and have learned the laws by which they are connected, that we have acquired any thing deserving the name of science.

To learn one thing at a time—to learn that thing thoroughly—and to learn its connections with other things are the golden steps that lead to the temple of knowledge.

It will be seen, in Lessons XVI. and XVII., Section VII., that UNITY has been *employed* to denote any number entering into an arithmetical question. This use of unity affords a powerful means of solving most questions which otherwise present great difficulties; and is, it is believed, a link of closer connection between the subjects of arithmetic and algebra, than has before been used.

It has been the author's aim, in the present work, to treat the subject of number in accordance with these principles, and to give to the whole a scientific form, and logical development. That he might not fail in so difficult and delicate an undertaking, he has defined all the terms, and given a full analysis of every process employed.

The work is complete in itself. It is a mental analysis of the science of numbers, designed to be accessible to the youngest pupils because of its simple gradations, and useful to the advanced pupil because of its scientific arrangement, its logical connections and its higher analysis of the properties and relations of numbers.

In the preparation of this work, many valuable suggestions and methods have been furnished by practical teachers. They were cheerfully offered and thankfully adopted.

FISHKILL LANDING, *February*, 1854.

INTELLECTUAL ARITHMETIC.

SECTION FIRST.

LESSON I.

Counting.

One,	*
Two,	* *
Three,	* * *
Four,	* * * *
Five,	* * * * *
Six,	* * * * * *
Seven,	* * * * * * *
Eight,	* * * * * * * *
Nine,	* * * * * * * * *
Ten,	* * * * * * * * * *
Eleven,	* * * * * * * * * * *
Twelve,	* * * * * * * * * * * *
Thirteen,	* * * * * * * * * * * * *
Fourteen,	* * * * * * * * * * * * * *
Fifteen,	* * * * * * * * * * * * * * *
Sixteen,	* * * * * * * * * * * * * * * *
Seventeen,	* * * * * * * * * * * * * * * * *
Eighteen,	* * * * * * * * * * * * * * * * * *
Nineteen,	* * * * * * * * * * * * * * * * * * *
Twenty,	* * * * * * * * * * * * * * * * * * * *

SUGGESTIONS.—There is but one simple idea in Arithmetic—it is the idea of the unit ONE.

Any collection of units is a number. Hence, every number is derived from one, and consequently has one for a base. Counting is merely naming numbers. In this lesson, the names of numbers are written opposite the collection.

How many units in four? In six apples, what is the unit? What is the unit in seven pears? How many units in twelve peaches? What is the unit that is counted in the lesson? How many stars in the 4th line? How many in the 14th?

From what are all numbers derived? What is the base of every number?

LESSON II.

Figures from One to Twenty.

1	. . . . *
2	. . . . * *
3	. . . . * * *
4	. . . . * * * *
5	. . . . * * * * *
6	. . . . * * * * * *
7	. . . . * * * * * * *
8	. . . . * * * * * * * *
9	. . . . * * * * * * * * *
10	. . . . * * * * * * * * * *
11	. . . . * * * * * * * * * * *
12	. . . . * * * * * * * * * * * *
13	. . . . * * * * * * * * * * * * *
14	. . . . * * * * * * * * * * * * * *
15	. . . . . . * * * * * * * * * * * * * * *
16	. . . . . * * * * * * * * * * * * * * * *
17	. . . . * * * * * * * * * * * * * * * * *
18	. . . * * * * * * * * * * * * * * * * * *
19	. . * * * * * * * * * * * * * * * * * * *
20	. *

Which figure stands for two? Which figure stands for four? Which figure stands for nine? Which stands for eight? What stands for ten? What stands for twelve? What stands for fourteen? What stands for sixteen? What stands for eighteen? What stands for twenty? What stands for seventeen? What stands for fifteen? What stands for nineteen? What stands for thirteen?

SUGGESTIONS.—This lesson is intended to teach that numbers may be expressed by figures, as well as by words. The teacher should explain to the pupil that the figure 2 and the word *two*, have the same meaning, and similarly for every figure and its corresponding word. Either the figure or the word, denotes as many units as its name points out.

LESSON III.

Figures from One to One Hundred.

Naught . .	0	Thirty-four .	34	Sixty-eight .	68
One . . .	1	Thirty-five .	35	Sixty-nine .	69
Two . . .	2	Thirty-six .	36	Seventy . .	70
Three . . .	3	Thirty-seven	37	Seventy-one .	71
Four . . .	4	Thirty-eight .	38	Seventy-two .	72
Five . . .	5	Thirty-nine .	39	Seventy-three	73
Six. . . .	6	Forty . . .	40	Seventy-four	74
Seven . . .	7	Forty-one .	41	Seventy-five .	75
Eight . . .	8	Forty-two .	42	Seventy-six .	76
Nine . . .	9	Forty-three .	43	Seventy-seven	77
Ten . . .	10	Forty-four .	44	Seventy-eight	78
Eleven . .	11	Forty-five .	45	Seventy-nine	79
Twelve . .	12	Forty-six. .	46	Eighty . .	80
Thirteen . .	13	Forty-seven .	47	Eighty-one .	81
Fourteen . .	14	Forty-eight .	48	Eighty-two .	82
Fifteen . .	15	Forty-nine .	49	Eighty-three	83
Sixteen . .	16	Fifty . . .	50	Eighty-four .	84
Seventeen .	17	Fifty-one. .	51	Eighty-five .	85
Eighteen . .	18	Fifty-two .	52	Eighty-six .	86
Nineteen . .	19	Fifty-three .	53	Eighty-seven	87
Twenty . .	20	Fifty-four .	54	Eighty-eight	88
Twenty-one .	21	Fifty-five .	55	Eighty-nine .	89
Twenty-two .	22	Fifty-six . .	56	Ninety . .	90
Twenty-three	23	Fifty-seven .	57	Ninety-one .	91
Twenty-four .	24	Fifty-eight .	58	Ninety-two .	92
Twenty-five .	25	Fifty-nine .	59	Ninety-three	93
Twenty-six .	26	Sixty . . .	60	Ninety-four .	94
Twenty-seven	27	Sixty-one .	61	Ninety-five .	95
Twenty-eight	28	Sixty-two .	62	Ninety-six .	96
Twenty-nine	29	Sixty-three .	63	Ninety-seven	97
Thirty . .	30	Sixty-four .	64	Ninety-eight	98
Thirty-one .	31	Sixty-five .	65	Ninety-nine .	99
Thirty-two .	32	Sixty-six . .	66	One hundred	100
Thirty-three .	33	Sixty-seven .	67	Two hundred	200

SUGGESTION.—This lesson is intended to teach that the words and the figures are simply two forms of language.

LESSON IV.

Roman Table.

I	One	XX . .	Twenty
II . . .	Two	XXI .	Twenty-one
III . . .	Three	XXX .	Thirty
IV . . .	Four	XL . .	Forty
V . . .	Five	L . .	Fifty
VI . . .	Six	LX . .	Sixty
VII . . .	Seven	LXX .	Seventy
VIII . . .	Eight	LXXX .	Eighty
IX . . .	Nine	XC . .	Ninety
X . . .	Ten	C . .	One hundred
XI . . .	Eleven	CC . .	Two hundred
XII . . .	Twelve	CCC . .	Three hundred
XIII . . .	Thirteen	CCCC .	Four hundred
XIV . .	Fourteen	D . .	Five hnndred
XV . . .	Fifteen	DC . .	Six hundred
XVI . .	Sixteen	DCC .	Seven hundred
XVII . .	Seventeen	DCCC .	Eight hundred
XVIII . .	Eighteen	DCCCC	Nine hundred
XIX . .	Nineteen	M . .	One thousand

This table is read, one I, one; two I's, two; three I's, three; IV, four, &c.

What stands for two? What stands for four? What stands for five? What stands for eight? What stands for ten? What stands for twenty? What stands for thirty? What stands for forty? What stands for fifty? What stands for sixty? What stands for seventy? What stands for eighty? What stands for ninety? What stands for one hundred? What stands for five hundred? What for one thousand?

SUGGESTION.—This lesson is intended to teach that numbers may be expressed by the Roman characters, as well as by words and by figures. Hence, there are three ways of expressing numbers, viz: by words, by figures, and by letters.

LESSON V.

Addition Table from 1 to 3 inclusive.

1. One and one, are how many*?
2. One and two, are how many?
3. One and three, are how many?
4. One and four, are how many?
5. One and five, are how many?
6. One and six, are how many?
7. One and seven, are how many?
8. One and eight, are how many?
9. One and nine, are how many?
10. One and ten, are how many?
11. Two and one, are how many?
12. Two and two, are how many?
13. Two and three, are how many?
14. Two and four, are how many?
15. Two and five, are how many?
16. Two and six, are how many?
17. Two and seven, are how many?
18. Two and eight, are how many?
19. Two and nine, are how many?
20. Two and ten, are how many?
21. Three and one, are how many?
22. Three and two, are how many?
23. Three and three, are how many?
24. Three and four, are how many?
25. Three and five, are how many?
26. Three and six, are how many?
27. Three and seven, are how many?
28. Three and eight, are how many?
29. Three and nine, are how many?
30. Three and ten, are how many?

---

* SUGGESTIONS.—The sum of two or more numbers contains as many units as there are in the numbers added. Thus, 2 is the sum of 1 and 1; 4 the sum of 2 and 2, or of 1 and 3.

ADDITION is the process of finding the sum of two or more numbers. The sign + (plus,) placed between two numbers signifies that they are to be added.

QUESTIONS.

1. How many fingers have you on one hand, not counting the thumb? How many on both hands?

2. Counting the thumb, how many have you on each hand? How many on both?

3. One and four are how many? One and five? One and nine? One and ten?

4. James has one apple and buys five: how many will he then have?

5. John has six apples and buys one: how many will he then have?

6. Charles has nine marbles and John gives him one: how many will he then have?

7. How many are 2 and 2? How many are 2 and 4? 2 and 3? 2 and 6?

8. How many are 2 and 7? How many are 2 and 8? How many are 2 and 9?

9. James has two tops and buys four: how many will he then have? Two and four are how many?

10. John has two apples and William gives him six: how many will he then have?

11. Bought two quills for two cents, and four quills for four cents: how many quills did I buy?

12. James bought 2 apples for two cents and 8 more for eight cents: how many did he buy in all? Three and six are how many?

13. If you buy two peaches for two cents and 9 peaches for nine cents, how many peaches do you buy? How much do you pay for them?

14. How many are 3 and 3? How many are 3 and 4?

15. John has three nuts in one hand and five in the other: how many in both? 3 and 8 are how many?

16. James has three pencils and John five: how many have both? 3 and six are how many?

LESSON VI.

Addition Table from 4 to 6 inclusive.

1. Four and one are how many ?
2. Four and two are how many ?
3. Four and three are how many ?
4. Four and four are how many ?
5. Four and five are how many ?
6. Four and six are how many ?
7. Four and seven are how many ?
8. Four and eight are how many ?
9. Four and nine are how many ?
10. Four and ten are how many ?
11. Five and one are how many ?
12. Five and two are how many ?
13. Five and three and how many ?
14. Five and four are how many ?
15. Five and five are how many ?
16. Five and six are how many ?
17. Five and seven are how many ?
18. Five and eight are how many ?
19. Five and nine are how many ?
20. Five and ten are how many ?
21. Six and one are how many ?
22. Six and two are how many ?
23. Six and three are how many ?
24. Six and four are how many ?
25. Six and five are how many ?
26. Six and six are how many ?
27. Six and seven are how many ?
28. Six and eight are how many ?
29. Six and nine are how many ?
30. Six and ten are how many ?

QUESTIONS.

1. John has four tops and Charles one : how many have both ?

2. William has four apples and James three: how many have both?

3. How many are 4 and 4? How many are 4 and 5?

4. John has four chestnuts in one hand and three in the other: how many has he in both?

5. Charles has four quills and John seven: how many have both? Four and 7 are how many?

6. John and James have each four tops: how many have both? Four and 9 are how many?

7. William has four birds in one cage and seven in another: how many in both?

8. Jane has four pins in her cushion and puts in six more: how many will she then have?

9. Mary has four needles and buys eight: how many will she then have? 4 and 10 are how many?

10. John buys three pears for four cents and six pears for eight cents: how many pears does he buy?

11. How many are 5 and 1? How many are 5 and 3? 5 and 8? 5 and 9? 5 and 10?

12. John has five marbles in one hand and eight in the other: how many in both?

13. Charles has five cents and his father gives him seven: how many has he then?

14. John has five apples and Reuben gives him nine: how many has he then?

15. Isaac buys five sheets of paper for five cents, and ten sheets more for ten cents: how many sheets does he buy?

16. If I buy five oranges for five cents, and six oranges for six cents, how many do I buy? Five and 5 are how many?

17. How many are 6 and 1? How many are 6 and 2? 6 and 6? 6 and 8? 6 and 10?

18. How many are 6 and 6? How many are 6 and 8? 6 and 9? 6 and 2? 6 and 10?

19. William carries six apples to school in his basket and Henry four: how many in both baskets?

20. John has six apples, and his sister Jane gives him five: how many has he then?

21. Charles has six apples and wins eight from John: how many has he then?

22. William buys three tops for six cents and eight tops for ten cents: how many tops does he buy?

23. James buys six eggs for six cents and eight eggs for nine cents: how many eggs does he buy?

24. Jane has 6 apples and Mary gives her 9: how many will she then have?

LESSON VII.

Addition Table from 7 to 9 inclusive.

1. Seven and one are how many?
2. Seven and two are how many?
3. Seven and three are how many?
4. Seven and four are how many?
5. Seven and five are how many?
6. Seven and six are how many?
7. Seven and seven are how many?
8. Seven and eight are how many?
9. Seven and nine are how many?
10. Seven and ten are how many?

11. Eight and one are how many?
12. Eight and two are how many?
13. Eight and three are how many?
14. Eight and four are how many?
15. Eight and five are how many?
16. Eight and six are how many?

17. Eight and seven are how many?
18. Eight and eight are how many?
19. Eight and nine are how many?
20. Eight and ten are how many?
21. Nine and one are how many?
22. Nine and two are how many?
23. Nine and three are how many?
24. Nine and four are how many?
25. Nine and five are how many?
26. Nine and six are how many?
27. Nine and seven are how many?
28. Nine and eight are how many?
29. Nine and nine are how many?
30. Nine and ten are how many?

QUESTIONS.

1. How many are 7 and 1? How many are 7 and 2? 7 and 5? 7 and 4? 7 and 6?

2. James has seven oranges in one basket and six in another: how many in both?

3. William has seven apples and John gives him nine: how many has he then?

4. A father has two sons and gives seven cents to each: how many cents does he give to both?

5. If Henry buys seven apples, and Mary gives him nine: how many will he then have?

6. If George buys seven quills at one time and 8 at another: how many does he buy in all?

7. William has 5 marbles and Henry gives him 8: how many will he then have?

8. How many are 8 and 2? How many are 8 and 4? 8 and 6? 8 and 5? 8 and 9?

9. A boy has eight marbles and gains five: how many has he then? 8 and 10 are how many?

10. If he has eight and gains nine, how many will he have? 6 and 4 and 5, are how many?

11. If George buys eight marbles for three cents and eight more for four cents, how many will he buy in all?

12. John has eight marbles and Charles gives him nine: how many has he then?

13. Eight and four are how many? Eight and seven how many? Eight and 6 are how many?

14. How many are 9 and 2? How many are 9 and 4? 1 and 3 and 6, are how many?

15. Charles has nine apples and buys five more: how many has he then? 9 and 9, are how many?

16. If he has nine and buys eight, how many will he have? 4 and 5 and 6, are how many?

17. Nine and seven are how many? 8 and 7?

18. If James buys nine oranges for nine cents and eight more for 9 cents, how many will he buy in all?

19. Six sheets of paper cost nine cents and 2 pencils cost one cent: what do the paper and pencils cost? 6 and 7 and 8, are how many?

20. How many are 10 and 2? How many are 10 and 3? 10 and 9? 8 and 7 and 6?

21. James has ten pencils and then buys eight: how many has he then? 9 and 4 and 3, are how many?

22. John gives ten chestnuts to Henry and nine to William: how many does he give to both?

23. James spends six cents for candy, four cents for gingerbread, and eight cents for nuts: how much does he spend in all? 1 and 2 and 4 and 10, are how many?

24. Nancy has ten pins on her cushion, and sticks nine more there: how many will she then have?

25. Jane has ten needles and Lucy gives her seven: how many will she then have?

26. Oliver buys ten oranges for twelve cents and ten more for eight cents: how many does he buy?

LESSON VIII.

Numbers added with reference to Ten.

1. Five and 5 are how many? 4 and 6? 3 and 7? 2 and 8? 1 and 9?*

2. Ten and 10 are how many? 15 and 5 are how many? 16 and 4? 17 and 3? 18 and 2? 19 and ? 14 and 6? 13 and 7? 12 and 8? 11 and 9?

3. Twenty and ten are how many? 25 and 5? 26 and 4? 27 and 3? 28 and 2? 29 and 1? 24 and 6? 23 and 7? 22 and 8? 21 and 9?

4. Thirty and 10 are how many? 35 and 5? 36 and 4? 37 and 3? 38 and 2? 39 and 1? 34 and 6? 33 and 7? 32 and 8? 31 and 9?

5. Forty and 10 are how many? 45 and 5? 46 and 4? 47 and 3? 48 and 2? 49 and 1? 44 and 6? 43 and 7? 42 and 8? 41 and 9?

6. Fifty and 10 are how many? 55 and 5? 56 and 4? 57 and 3? 58 and 2? 59 and 1? 54 and 6? 53 and 7? 52 and 8? 51 and 9?

7. Sixty and 10 are how many? 65 and 5? 66 and 4? 67 and 3? 68 and 2? 69 and 1? 64 and 6? 63 and 7? 62 and 8? 61 and 9?

8. Seventy and 10 are how many? 75 and 5? 76 and 4? 77 and 3? 78 and 2? 79 and 1?

9. Eighty and 10 are how many? 85 and 5? 89 and 1? 87 and 3? 4 and 86? 2 and 88? 6 and 84? 7 and 83? 9 and 81? 8 and 82?

10. Ninety and ten are how many? 99 and 1? 92 and 8? 7 and 93? 3 and 97? 5 and 95? 6 and 94? 9 and 91? 98 and 2? 4 and 96?

---

*SUGGESTIONS.—The first part of this lesson points out to the pupil the method of adding, by considering what numbers make exact tens. This is very important, and should be much dwelt [illegible]

The second part shows that the first figure of a sum is always derived from the units.

The questions of the lesson should first be put in the order in which they are written, and then promiscuously, until thoroughly learned.

11. Two and two are how many? 12 and 2? 22 and 2? 32 and 2? 42 and 2? 52 and 2? 62 and 2? 72 and 2? 82 and 2? 92 and 2? 94 and 2? 96 and 2? 98 and 2?

12. Three and 3 are how many? 13 and 3? 23 and 3? 33 and 3? 43 and 3? 53 and 3? 63 and 3? 73 and 3? 83 and 3? 93 and 3? 96 and 4?

13. Four and 4 are how many? 4 and 14? 24 and 4? 34 and 4? 44 and 4? 54 and 4? 64 and 4? 74 and 4? 84 and 4? 94 and 4? 98 and 2?

14. Five and 5 are how many? 15 and 5? 25 and 5? 35 and 5? 45 and 5? 55 and 5? 65 and 5? 75 and 5? 85 and 5? 95 and 5?

15. Six and 6 are how many? 16 and 6? 26 and 6? 36 and 6? 46 and 6? 66 and 6? 76 and 6? 86 and 6? 96 and 6?

16. Seven and 7 are how many? 17 and 7? 27 and 7? 37 and 7? 47 and 7? 57 and 7? 67 and 7? 77 and 7? 87 and 7? 97 and 7?

17. Eight and 8 are how many? 18 and 8? 28 nd 8? 38 and 8? 48 and 8? 58 and 8? 68 and eight? 78 and 8? 88 and 8? 98 and 8?

18. Nine and 9 are how many? 19 and 9? 29 and 9? 39 and 9? 49 and 9? 59 and 9? 69 and 9? 79 and 9? 89 and 9? 99 and 9?

LESSON IX

Showing the formation of Numbers from 11 *to* 100.

1. Eleven and 1 are how many? Eleven and 2? Eleven and 3? Eleven and 4? Eleven and 5? Eleven and 6? Eleven and 7? Eleven and 8? Eleven and 9? Eleven and 10? Eleven and 11?*

* SUGGESTION.—This lesson indicates how all the numbers may be formed from 11 to include one hundred and nine. After the questions have been put in the order in which they are written they should be put promiscuously.

2 Twenty-two and 1 are how many? Twenty-two and 2? Twenty-two and 3? Twenty-two and 4? Twenty-two and 5? Twenty-two and 6? Twenty-two and 7? Twenty-two and 8? Twenty-two and 9? Twenty-two and 10? Twenty-two and 11?

3. Thirty-three and 1 are how many? Thirty-three and 2? Thirty-three and 3? Thirty-three and 4? Thirty-three and 5? Thirty-three and 6? Thirty-three and 7? Thirty-three and 8? Thirty-three and 9? Thirty-three and 10? Thirty-three and 11?

4. Forty-four and 1 are how many? Forty-four and 2? Forty-four and 3? Forty-four and 4? Forty-four and 5? Forty-four and 6? Forty-four and 7? Forty-four and 8? Forty-four and 9? Forty-four and 10? Forty-four and 11?

5. Fifty-five and 1 are how many? Fifty-five and 2? Fifty-five and 3? Fifty-five and 4? Fifty-five and 5? Fifty-five and 6? Fifty-five and 7? Fifty-five and 8? Fifty-five and 9? Fifty-five and 10? Fifty-five and 11?

6. Sixty-six and 1 are how many? Sixty-six and 2? Sixty-six and 3? Sixty-six and 4? Sixty-six and 5? Sixty-six and 6? Sixty-six and 7? Sixty-six and 8? Sixty-six and 9? Sixty-six and 10? Sixty-six and 11?

7. Seventy-seven and 1 are how many? Seventy-seven and 2? Seventy-seven and 3? Seventy-seven and 4? Seventy-seven and 5? Seventy-seven and 6? Seventy-seven and 7? Seventy-seven and 8? Seventy-seven and 9? Seventy-seven and 10? Seventy-seven and 11?

8. Eighty-eight and 1 are how many? Eighty-eight and 2? Eighty-eight and 3? Eighty-eight and 4? Eighty-eight and 5? Eighty-eight and 6? Eighty-eight and 7? Eighty-eight and 8? Eighty

eight and 9? Eighty-eight and 10? Eighty-eight and 11?

9. Ninety-nine and 1 are how many? Ninety-nine and 2? Ninety-nine and 3? Ninety-nine and 4? Ninety-nine and 5? Ninety-nine and 6? Ninety-nine and 7? Ninety-nine and 8? Ninety-nine and 9? Ninety-nine and 10?

LESSON X.

Practical Questions.

1. Let each of the following combinations be given as a separate example. How many are

10 and 20 and 4?	40 and 50 and 6?
10 and 30 and 9?	6 and 12 and 30?
10 and 40 and 6?	7 and 15 and 70?
10 and 50 and 3?	9 and 14 and 60?
20 and 30 and [illegible]?	13 and 7 and 14?
15 and 20 and 6?	19 and 11 and 16?
25 and 15 and 4?	21 and 9 and 13?
35 and 12 and 3?	30 and 40 and 10?
40 and 60 and 9?	36 and 4 and 19?
8 and 20 and 10?	38 and 12 and 16?

2. Jane has 13 pins in her cushion and Mary 27: how many pins have both?

3. John has a number of pears: he gives 8 to William, 12 to Charles, 9 to James and has 1 left: how many had he at first?

4. There are 4 bags of coffee: the first contains 16 pounds, the second 14 pounds, the third 7 pounds, and the fourth 3 pounds: how many pounds in all the bags?

5. A farmer has 4 pastures containing sheep. In the first there are 3 sheep, in the second there are 6, in

the third there are 7, and in the fourth there are 8: how many are there in the four pastures?

6. James gave 18 cents for a squirrel, 82 cents for a cage, and 15 cents for nuts: how many cents did he pay in all?

7. A man bought a cow for 25 dollars, a calf for 5 dollars, 3 lambs for 8 dollars, and a pig for 2 dollars: what did he pay for all?

8. How many are 1 and 2 and 4 and 14 and 9?

9. How many are 3 and 4 and 16 and 5 and 4 and 5?

10. How many are 4 and 14 and 16 and 6 and 7 and 8?

11. How many are 15 and 13 and 12 and 4 and 9?

12. How many are 9 and 11 and 14 and 16 and 17?

13. How many are 1 and 2 and 4 and 3 and 6?

14. How many are 2 and 2 and 4 and 3 and 5?

15. How many are 6 and 4 and 4 and 3 and 3?

16. How many are 6 and 4 and 3 and 6 and 5?

17. How many are 7 and 7 and 4 and 2 and 6?

18. How many are 9 and 2 and 8 and 7 and 5 and 8?

19. A lady bought some tape for 10 cents, some pins for 18 cents, a comb for 22 cents, and a pair of scissors for 30 cents: how much did she pay in all?

20. A farmer has 15 sheep in one lot, 25 in another, and 30 in his barn-yard: how many has he in all?

21. A merchant buys 26 barrels of flour of one miller, 30 of another, 14 of another, and 36 of another: how many barrels does he buy in all?

22. A man bought a horse, saddle and bridle; for the horse he gave 75 dollars, for the saddle 25 dollars, and 7 dollars for the bridle: what did they all cost him?

23. A drover bought 12 sheep of one farmer, 30 of another, 18 of another, and 25 of another: how many did he buy in all?

24. James gave nine cents to a beggar woman, 11 cents to a beggar man, and 8 cents to a beggar girl: how much did he give in all?

25. If John has 14 cents in one pocket, 10 cents in another, 6 cents in his purse, and 8 cents in his hand, how many cents has he in all?

26. Charles has 8 cents, William 18, Robert 4, and Samuel 9: how many cents have they all?

27. If Lucius gives 36 cents for a pen-knife, 8 cents for paper, 6 cents for quills, and 7 cents for wafers, how much does he pay in all?

28. James buys 9 sticks of white candy, 9 sticks of red candy, 2 of brown candy, and 8 of yellow candy: how many sticks does he buy in all?

29. If James gave 18 cents for the white candy, 18 cents for the red candy, 4 cents for the brown candy, and 16 cents for the yellow candy, how much did he pay in all?

30. Jane pays 18 cents for a slate, 12 cents for quills, 11 cents for paper, and 15 cents for pencils: what does she pay in all?

31. A grocer purchases 6 barrels of flour for 30 dollars, a load of hay for 12 dollars, and 10 bushels of oats for 15 dollars: how much did he pay in all?

32. A tailor paid 15 dollars for a piece of cloth, for a coat, 4 dollars for the lining, 2 dollars for the buttons, and charged 9 dollars for making: what was the cost of the coat?

SECTION SECOND.

LESSON I.

Subtraction.

1. One and 1 are 2: if we take 1 from 2 what remains*?

2. One and 2 are 3: if we take 1 from 3, what remains? If we take 2 from 3, what remains?

3. One and 3 are 4: if we take 1 from 4, what remains? If we take 3 from 4, what remains? If we take 2 from 4, what remains?

4. One and 4 are 5: if we take 1 from 5, what remains? 5 less 4, are how many? 5 less 2, are how many? 5 less 3, are how many?

5. One and 5 are 6: 6 less 5, are how many? 6 less 1, are how many? 6 less 2, are how many? 6 less 3, are how many? 6 less 4, are how many?

6. Seven less 1 are how many? 7 less 2, are how many? 7 less 4, are how many? 7 less 5, are how many? 7 less 6, are how many?

---

* SUGGESTIONS.—The pupil first gets the idea of *more* by addition. He adds two numbers together and finds that their sum is greater than either of them. He next sees that if one of them be taken away, the other will be left: hence,

The difference between two numbers is such a number as added to the less will give the greater.

SUBTRACTION is the process of finding the difierence between two numbers.

Teach the pupil that every question in Subtraction requires him to find such a number as added to the less will give the greater.

What is the difference between 6 and 2? Why is 4 the difference between 6 and 2? Because 4 added to 2 gives 6.

The sign — (minus), placed between two numbers signifies that the number after it is to be taken from the number before it.

7. Eight less 1, are how many? 8 less 3, are how many? 8 less 5, are how many? 8 less 7, are how many?

8. Nine less 1, are how many? 9 less 3, are how many? 9 less 5, are how many?

9. Eight and 2, are how many? 10 less 2, are how many? 10 less 8, are how many?

10. Six and 4, are how many? 10 less 6, are how many? 11 less 5, are how many?

11. William has three apples and gives them all to James, how many has he left? 3 less 3, what remains?

12. William has six apples and gives three to James: how many has he left? 6 less 3, are how many?

13. 7 less 3, are how many?

14. 9 less 3, are how many?

15. 10 less 3, are how many?

16. 14 less 3, are how many?

17. 4 less 4, what remains?

18. 8 less 4, are how many?

19. Fourteen less 4, are how many?

20. Henry has five pears in a basket and gives them all to his sister: how many has he left? 5 less 5 what remains?

21. 8 less 5, leaves how many?

22. 9 less 5, leaves how many?

23. 11 less 5, leaves how many?

24. James has six squirrels in a cage, and takes them all out: how many will be left? 6 from 6, leaves how many?

25. 9 less 6, leaves how many?

26. 8 less 6, are how many?

27. 16 less 6, are how many?

28. Mary has seven pins in her cushion and takes them all out: how many are left? 7 less 7, what remains?

29. 12 less 7, are how many?
30. 15 less 7, are how many?
31. Reuben has eight plums and gives them all to John: how many has he left? 8 less 8, what remains?
32. 10 less 8, are how many?
33. 14 less 8, are haw many?
34. 18 less 8, are how many?
35. There are nine chairs in a room, and Mary takes them all out: how many are left? 9 less 9, what remains?
36. 12 less 9, are how many?
37. 15 less 9, are how many?
38. 19 less 9, are how many?
39. 16 less 6, are how many?
40. 15 less 10, are how many?
41. 14 less 4, are how many?
42. 25 less 5, are how many?
43. 36 less 16, are how many?
44. 25 less 12, are how many?
45. 30 less 7, are how many?
46. 20 less 14, are how many?
47. 30 less 12, are how many?
48. 27 less 7, are how many?
49. 29 less 10, are how many?
50. 39 less 20, are how many?
51. 42 less 12, are how many?
52. 19 less 7, are how many?
53. 50 less 20, are how many?
54. 14 less 8, are how many?
55. 15 less 3, are how many?
56. 24 less 4, are how many?
57. 16 less 5, are how many?
58. 17 less 8, are how many?
59. 19 less 7, are how many?
60. 19 less 9, are how many?

61. 29 less 8, are how many?
62. 34 less 3, are how many?
63. 35 less 6, are how many?
64. 50 less 8, are how many?
65. 57 less 6, are how many?
66. 59 less 5, are how many?
67. 53 less 7, are how many?
68. 60 less 20, are how many?
69. 65 less 15, are how many?
70. 67 less 10, are how many?
71. 67 less 9, are how many?
72. 70 less 5, are how many?
73. 74 less 3, are how many?
74. 78 less 8, are how many?
75. 79 less 10, are how many?

QUESTIONS.

1. There are nineteen peach-trees in an orchard, and six of them are blown down in a storm: how many are left standing?

2. Laura has twenty-five cents, and buys an arithmetic for eighteen cents: how much money will she have left?

3. There are thirty-four pears in a basket, and nine of them are taken out: how many are left?

4. There are sixty-five pigeons in a flock, and John fires at them and kills nine: how many are left?

5. There are fifty-four sheep in a fold, and a wolf breaks in and kills seven: how many are left?

6. There are forty-nine scholars in a school, and ten of them are girls: how many boys are there?

7. In another school there are twenty scholars, and nine are boys: how many girls are there?

8. In Elizabeth's flower-bed there are thirty beautiful lilies, and John breaks off seven of them: how many are unbroken?

9. James has thirty-seven cents: he spends six for candy, eight for a pencil, and twelve for a pen-knife: how many has he left?

10. John has twenty-five cents, and spends six cents for a top, nine cents for a pencil, and two cents for a peach: how much has he left?

11. Wiliam has 37 cents, he buys a top for 16 cents, and 8 marbles for 2 cents: how much has he left?

12. A boy has 40 peaches: he gives 24 to Lucy and 9 to Elizabeth: how many has he left?

13. James received a premium worth 56 cents; Jane received one worth 30 cents: what was the difference of their values?

14. Charles has 49 cents, and buys a book which costs him 29 cents: how much has he left?

15. A butcher buys 39 sheep, and kills 17: how many are left alive?

16. A grocer has a tub of butter containing 45 pounds; he sells 20 pounds to Mr. Wilson, and 15 pounds to Mr. Jones: how much is left?

17. How many are 55 less 17?

18. How many are 77 less 19?

19. Four men bought a horse for 56 dollars; the first paid 16 dollars, the second 20 dollars, and the third 14 dollars: what did the fourth pay?

20. Charles bought a penknife for 48 cents, and a top for 25 cents: how much more did he pay for the knife than for the top?

LESSON II.

Questions involving Addtition and Substraction.

1. Six and 4 and 3 and 5, less 4, are how many?
2. Eight and 9 and 4 and 2, less 3, are how many?
3. Seven and 6 and 5 and 8, less 5, are how many?

4. Nine and 2 and 3 and 7, and 1, less 6, are how many?

5. Thirty and 5 and 8 and 15, less 10, are how many?

6. Twenty-one and 6 and 13, less 12, are how many?

7. Forty-five and 15 and 12, less 9, are how many?

8. Sixty-nine and 11 and 5 and 2, less 8, are how many?

9. Seventy-five and 5 and 6 and 12, less 8, are how many?

10. Forty-five and 8 and 4 and 3 and 6 and 7 and 9, less 12, are how many?

11. James has 40 cents, and pays 12 cents for a whistle and 25 cents for a knife: how much has he left?

12. A man bought a calf for 5 dollars, a sheep for 4 dollars, and a pig for 2 dollars; also a cow for 25 dollars: what did he pay in all, and how much more for the cow than for the other animals?

13. James has 26 nuts in one pocket, and 14 less in the other pocket: how many has he in both?

14. A man has 50 dollars; he pays 26 for a coat, 8 dollars for a pair of pantaloons, and 4 dollars for a vest: how much has he left?

15. A school-boy pays 56 cents for an Atlas, 30 cents for an Arithmetic, and 24 cents for a slate: what did he pay for all, and how much more for the Atlas than for the Arithmetic and slate?

16. Eighty-five and 5 and 9 and 3 and 12, less 8, less 3, are how many?

17. Fifty-nine and 5 and 9 and 8 and 7 and 6, less 8, are how many?

18. Forty-seven and 9 and 6 and 4 and 5 and 3, less 12, are how many?

19. Seventy-two and 15 and 6, less 11, are how many?

20. Thirty eight and 4 and 9 and 7 and 6 and 2, less 12, are how many?

21. William had 16 marbles, James gave him 7, John gave him 8, and Reuben gave him enough to make his number, 40: how many marbles did Reuben give John?

22. A father gave seventeen cents to Lucy, 13 to Mary, and 4 to Jane, and then took back 9 from Lucy: how many had they left?

23. A man travelled 49 miles in three days; the first day he travelled 16 miles, and the second day 13 miles: how far did he travel the third day?

24. A tailor has a piece of cloth containing 39 yards; he sold 14 yards to one man, 13 to another, and made a coat which took two yards: how many yards were left?

25. A merchant bought some coffee, for which he paid 25 dollars, some sugar, for which he paid 12 dollars, and some tea, for which he paid 11 dollars; he sold the whole for 56 dollars: what did he gain?

26. A tailor bought a piece of cloth for 45 dollars; he made it into a coat and pantaloons; he paid 10 dollars for making, and then sold them for 60 dollars: did he make or lose, and how much?

27. James has 49 peaches; he gives 15 to Robert, 13 to John, and 9 to William: how many has he left?

28. Charles has some pears, and gives 12 to James, 11 to Henry, 13 to Reuben, and 8 to Elisha; when he finds that he has 9 left: how many had he at first?

29. Lucy has 12 pins on one cushion, and 15 on another: if she takes off 9, how many will she have left?

30. A man travelled 5 miles before breakfast, 19 miles between breakfast and dinner, and then travelled back 12 miles: how far was he from the place of starting?

31. A cow has two calves, the first is worth 3 dollars, the second 4 dollars, and the cow is worth 25 dollars: how much more is the cow worth than the two calves, and what are they all worth?

32. A grocer buys some lemons for 15 dollars, some oranges for 25 dollars, and then sold the whole for 56 dollars: how much did he make?

33. Jane has 32 rose-buds on one bush, and 16 on another, and 38 only blossom: how many buds did not flower?

34. A man owes 55 dollars; at one time he pays 24 dollars, at another 17 dollars, and then 14 dollars: how much does he then owe?

35. William went after chestnuts; he put 26 in one pocket, 15 in another, and 16 in a third: he lost 9 out of the first pocket, 3 from the second, and 6 from the third: how many had he left?

36. Twenty-nine plus 6 plus 8 plus 4, less 12, less 11, are how many?

37. Mr. Jones owes his baker 17 dollars, his grocer 16 dollars, and his tailor 27 dollars: how much does he owe in all, and how much more to his baker and grocer than to his tailor?

38. A farmer has 30 sheep in one lot, 25 in another, and 16 in a third: 9 sheep escape from the first lot, 12 from the second, and 9 from the third: how many sheep had he in all, how many escaped, and how many were left in the fields?

SECTION THIRD.

LESSON I.

Multipliers from 1 to 4 inclusive.

Once 1 is how many ?*	Once 7 is how many ?
Once 2 is how many ?	Once 8 is how many ?
Once 3 is how many ?	Once 9 is how many ?
Once 4 is how many ?	Once 10 is how many ?
Once 5 is how many ?	Once 11 is how many ?
Once 6 is how many ?	Once 12 is how many ?
2 times 1 are how many ?	2 times 7 are how many ?
2 times 2 are how many ?	2 times 8 are how many ?
2 times 3 are how many ?	2 times 9 are how many ?
2 times 4 are how many ?	2 times 10 are how many ?
2 times 5 are how many ?	2 times 11 are how many ?
2 times 6 are how many ?	2 times 12 are how many ?
3 times 1 are how many ?	3 times 7 are how many ?
3 times 2 are how many ?	3 times 8 are how many ?
3 times 3 are how many ?	3 times 9 are how many ?
3 times 4 are how many ?	3 times 10 are how many ?
3 times 5 are how many ?	3 times 11 are how many ?
3 times 6 are how many ?	3 times 12 are how many ?
4 times 1 are how many ?	4 times 7 are how many ?
4 times 2 are how many ?	4 times 8 are how many ?
4 times 3 are how many ?	4 times 9 are how many ?
4 times 4 are how many ?	4 times 10 are how many ?
4 times 5 are how many ?	4 times 11 are how many ?
4 times 6 are how many ?	4 times 12 are how many ?

* MULTIPLICATION *is a short process of taking one number as many times as there are units in another.*

The number to be taken is called the *multiplicand.* The number denoting how many times the multiplicand is to be taken, is called the *multiplier.*

The result, or answer, is called the product.

The multiplicand and multiplier are called *factors* or *producers* of the product.

1. If James buys 2 oranges at 3 cents apiece, what do they cost?*

2. What will 4 peaches cost at 1 cent each?

3. What will 3 oranges cost at 3 cents apiece?

4. What will 4 lemons cost at 4 cents apiece?

5. What will 3 pounds of raisins cost at 12 cents a pound?

6. What will be the cost of 4 tops at 12 cents apiece?

7. What will be the cost of 2 melons at 11 cents apiece?

LESSON II.

Multipliers from 5 to 8 inclusive.

5 times 1 are how many?	5 times 7 are how many?
5 times 2 are how many?	5 times 8 are how many?
5 times 3 are how many?	5 times 9 are how many?
5 times 4 are how many?	5 times 10 are how many?
5 times 5 are how many?	5 times 11 are how many?
5 times 6 are how many?	5 times 12 are how many?
6 times 1 are how many?	6 times 7 are how many?
6 times 2 are how many?	6 times 8 are how many?
6 times 3 are how many?	6 times 9 are how many?
6 times 4 are how many?	6 times 10 are how many?
6 times 5 are how many?	6 times 11 are how many?
6 times 6 are how many?	6 times 12 are how many?
7 times 1 are how many?	7 times 7 are how many?
7 times 2 are how many?	7 times 8 are how many?
7 times 3 are how many?	7 times 9 are how many?
7 times 4 are how many?	7 times 10 are how many?
7 times 5 are how many?	7 times 11 are how many?
7 times 6 are how many?	7 times 12 are how many?

* ANALYSIS.—Two oranges will cost two times as much as one orange. Since 1 orange costs 3 cents 2 oranges will cost two times 3 cents, which are 6 cents: therefore, 2 oranges at 3 cents apiece, will cost 6 cents.

8 times 1 are how many?	8 times 7 are how many?
8 times 2 are how many?	8 times 8 are how many?
8 times 3 are how many?	8 times 9 are how many?
8 times 4 are how many?	8 times 10 are how many?
8 times 5 are how many?	8 times 11 are how many?
8 times 6 are how many?	8 times 12 are how many?

1. If William buys 5 pine-apples at 4 cents each, what do they cost him?

2. What is the cost of 6 barrels of flour at 5 dollars a barrel?

3. What is the cost of 7 yards of cloth at 9 dollars a yard?

4. What is the cost of 8 barrels of fish at 8 dollars a barrel?

5. What is the cost of 6 pounds of candles at 9 cents a pound?

6. What is the cost of 8 pounds of raisins at 11 cents a pound?

7. What is the cost of 7 pounds of sugar at 8 cents a pound?

8. What is the cost of 5 pounds of beef at 11 cents a pound?

LESSON III.

Multipliers from 9 *to* 12 *inclusive.*

9 times 1 are how many?	9 times 7 are how many?
9 times 2 are how many?	9 times 8 are how many?
9 times 3 are how many?	9 times 9 are how many?
9 times 4 are how many?	9 times 10 are how many?
9 times 5 are how many?	9 times 11 are how many?
9 times 6 are how many?	9 times 12 are how many?

NOTE.—The sign of multiplication (×), placed between two or more numbers, signifies that they are to be multiplied together.

10 times 1 are how many?	10 times 7 are how many?
10 times 2 are how many?	10 times 8 are how many?
10 times 3 are how many?	10 times 9 are how many?
10 times 4 are how many?	10 times 10 are how many?
10 times 5 are how many?	10 times 11 are how many?
10 times 6 are how many?	10 times 12 are how many?
11 times 1 are how many?	11 times 7 are how many?
11 times 2 are how many?	11 times 8 are how many?
11 times 3 are how many?	11 times 9 are how many?
11 times 4 are how many?	11 times 10 are how many?
11 times 5 are how many?	11 times 11 are how many?
11 times 6 are how many?	11 times 12 are how many?
12 times 1 are how many?	12 times 7 are how many?
12 times 2 are how many?	12 times 8 are how many?
12 times 3 are how many?	12 times 9 are how many?
12 times 4 are how many?	12 times 10 are how many?
12 times 5 are how many?	12 times 11 are how many?
12 times 6 are how many?	12 times 12 are how many?

1. If James buys 9 lemons at 2 cents each, what do they cost him?

2. What will 11 yards of calico cost at 11 cents a yard?

3. What will 12 dozen of apples cost at 12 cents a dozen?

4. What will 9 pine-apples cost at 11 cents apiece?

5. What will 12 yards of cloth cost at 9 dollars a yard?

6. What will 9 pumpkins cost at 12 cents apiece?

7. What will be the cost of 11 pairs of boots at 5 dollars a pair?

8. What will be the cost of 12 loaves of bread at 11 cents a loaf?

9. What will 9 slates cost at 11 cents apiece?

10. What will be the cost of 11 yards of broad cloth at 12 dollars a yard?

LESSON IV.

QUESTIONS.

1. What is the product of 13 taken 2 times?*
2. What is the product of 14 taken 3 times?
3. What is the product of 15 taken 5 times?
4. What is the product of 18 taken 6 times? 9 times? 7 times?
5. What is the product of 16 taken 4 times? 5 times? 8 times?
6. If James reads 16 verses of the bible a day, how many verses will he read in a week?
7. How many are 7 times 15? Which the multiplicand? Which the multiplier? Which the product?
8. What will 9 pounds of butter cost at 19 cents a pound?
9. If a steamboat can go 16 miles an hour, how far can it go in 11 hours?
10. How many are 2 times 20? 4 times 20? 5 times 20? 7 times 20? 9 times 20? 11 times 20? 12 times 20?
11. If 1 barrel of flour cost 9 dollars, what will 12 barrels cost?
12. If 1 barrel of fish cost 7 dollars, what will be the cost of 9 barrels?
13. If 1 yard of cloth cost 6 dollars, what will be the cost of 8 yards?
14. If 1 yard of cloth costs 5 dollars, what will 15 yards cost?
15. If a man on horseback can ride 8 miles in 1 hour, how far can he ride in 12 hours?

---

* ANALYSIS.—Thirteen is made up of 1 ten and 3 units. Then 1 ten multiplied by 2 gives 2 tens, or 20; and 3 units multiplied by 2 gives 6 units, which being added to 20 gives 26 for the product. Let each question be analyzed in a similar manner.

16. How many are 2 times 18? 3 times 18? 5 times?

17. How many are 5 times 17? 3 times 18? 4 times?

18. If 1 cow cost 20 dollars, how much will 7 cows cost? How much will 9 cost? How much will 12 cost?

19. If a man travels 19 miles a day, how far will he travel in 9 days? In 10 days? In 11 days? In 12 days?

20. If a pound of raisins cost 15 cents, how much will 5 pounds cost? 6 pounds? 7 pounds? 11 pounds?

21. If a man digs 18 bushels of potatoes in 1 day, how many would he dig in 7 days? In 10 days? In 12 days?

22. If a man eat 15 ounces of bread in a day, how many ounces will he eat in a week? How many in 9 days? In 11 days.

LESSON V.

Addition.—Subtraction.—Multiplication.

1. What will be the cost of 5 oranges at 3 cents apiece?

2. What will 7 books cost at 8 cents apiece? At 9 cents? At 15 cents? At 20 cents?

3. What will be the cost of 6 loaves of bread at 9 cents a loaf? Of 8 loaves? Of 10 loaves? Of 11 loaves? Of 12 loaves?

4. What will be the cost of 4 hats at 7 dollars apiece, and the cost of 2 coats at 12 dollars apiece?

5. What will be the cost of 12 chickens at 12 cent apiece? Of 11? Of 9? Of 7? Of 6? Of 5?

6. What will be the difference in cost between 2 [illegible]cils at 9 cents apiece, and 3 tops at 4 cents apiece? [illegible] they all cost?

7. What will be the cost of 8 yards of ribbon at 8 cents a yard? Of 5 yards? Of 6 yards? Of 9 yards? Of 12 yards?

8. What is the difference between the cost of 8 yards of cloth at 5 dollars a yard, and 12 yards at 2 dollars a yard? What would be the cost of the whole?

9. James has 25 marbles and gives 9 to William, then Charles gives him as many as he has left: how many more than he had at first will he then have?

10. What will be the whole cost of 8 apples at 1 cent apiece, and of 4 pears at 2 cents apiece?

11. What will be the whole cost of 6 oranges at 4 cents apiece, and of 2 lemons at 3 cents apiece?

12. What will be the whole cost of 6 lemons at 2 cents apiece, and of 2 apples at 1 cent apiece?

13. What will be the whole cost of 8 quills at 2 cents apiece, and of 12 sheets of paper at 1 cent each?

14. What will be the whole cost of 6 spelling books at 8 cents apiece, and of 5 slates at 10 cents apiece?

15. What will be the difference of the cost of 6 yards of cloth at 5 dollars a yard, and of 4 yards of cloth at 6 dollars a yard?

16. What will be the whole cost of ten sticks of candy at 2 cents apiece, and of 4 pounds of raisins at 11 cents a pound?

17. A farmer bought 9 sheep at three dollars apiece, and two calves at 4 dollars apiece: how much more did he pay for the sheep than for the calves?

18. A farmer bought 4 cows at 20 dollars apiece, and 12 sheep at 4 dollars apiece: what did the whole cost him?

19. Fifteen plus 8 plus 9, less 5, are how many?

20. Forty plus 7, less 6, plus 8 plus 9, are how many?

21. Twenty-five plus 6 plus 8, less 3, plus 1, are how many?

22. Fifty plus 6 plus 9, less 8, less 3, less 4, are how many?

23. Sixty plus 8 plus 7 plus 1, less 6, are how many?

24. Thirty less 7 less 6, plus 8, plus 9, are how many?

25. Mary has 6 rose-bushes, and 9 buds on each; also 3 geraniums with 8 buds on each: how many buds are there in all?

26. A family consumes 12 pounds of meat in a day. Beef is 12 cents a pound and mutton 9: how much will they save by using mutton instead of beef?

27. James bought 4 oranges at 3 cents apiece, and 2 quarts of chestnuts at 9 cents a quart: what did he pay in all?

28. James and John start from the same place, and run in opposite directions. James runs 30 rods and John 20: how far are they apart?

How far would they be apart if they had run the same way?

29. Two men start from the same place and walk in opposite directions: the first walks 4 hours and goes 8 miles an hour, the second the same time and goes but 2 miles an hour: how far will they be apart?

30. Two men start from the same place and walk the same way; the first walks 4 miles an hour, the second 3: how far will they be apart at the end of the first hour? How far at the end of the second? The third?

31. A drover bought 3 sheep at 4 dollars apiece, and 4 lambs at 2 dollars apiece: he gave in payment 2 calves at 5 dollars apiece, and the rest in cash. How much money did he pay?

32. A jeweller bought a watch for 55 dollars, a

chain for 15 dollars, and a seal for 8 dollars; and then sold the whole for 80 dollars: did he make or lose, and how much?

33. Four boys bought a foot-ball for 75 cents. John paid 20 cents, James 33 cents, and William 18 cents: how much did Reuben pay?

34. A grocer bought a hogshead of sugar for 65 dollars, and a hogshead of molasses for 55 dollars; he sold them both for 125: did he make or lose, and how much?

35. A farmer bought 4 sheep for 15 dollars, and 6 lambs for 10 dollars; he sold the whole for 30 dollars: did he make or lose, and how much?

36. A tailor has a piece of cloth containing 25 yards, and a second piece containing 35 yards; he cut 8 yards from the first piece, and 12 from the second: how many yards of cloth has he left?

37. A merchant has a piece of cloth, for which he paid 36 dollars; he wishes to sell it so as to make a profit of 12 dollars: what must he ask for it?

38. A farmer has 150 sheep in three fields. In the first he has 45, in the second 55: how many has he in the third?

39. There is an orchard with ten full rows of trees and 7 trees in each row; besides which there are 3 broken rows with 4 trees in each row: how many trees are there in the orchard? How many more in the whole than in the broken rows?

40. An orchard contains 20 apple trees, 10 cherry trees, and 15 plum trees: how many more apple and cherry trees than plum trees?

41. James worked 4 days in a week for 12 cents a day, and William 2 days for 4 cents a day: how much more did James earn than William?

42. A butcher bought 16 sheep of one man, 14 of another, and 25 of a third; he then killed 9, afterwards he killed 21: how many had he left?

43. A man earned 16 dollars a month for 6 months and his son James 7 dollars a month for the same time: how much more did the father earn than the son?

LESSON VI

Factoring Numbers.

1. Four are how many times 2? What are the factors of 4?*

2. Six are how many times 3? How many times 2? What are the factors of 6?

3. Eight are how many times 4? How many times 2? What are the factors of 8?

4. Nine are how many times 3? What are the factors of 9?

5. Ten are how many times 5? How many times 2? What are the factors of 10?

6. Twelve are how many times 6? How many times 2? How many times 4? How many times 3? What are the factors of 12?

7. Fourteen are how many times 7? How many times 2? What are the factors of 14?

8. Fifteen are how many times 5? How many times 3? What are the factors of 15?

9. Sixteen are how many times 8? How many times 2? How many times 4? What are the factors of 16?

10. Eighteen are how many times 9? How many times 2? How many times 6? How many times 3? What are the factors of 18?

---

*Suggestion.—In each of the following questions, two numbers are named, and it is required to find a third which multiplied by the second, will give a product equal to the first. The second and third numbers are called FACTORS of the first.

Let the pupil point out the factors in every example.

4*

11. Twenty are how many times 2? How many times 10? How many times 5? How many times 4? What are the factors of 20?

12. Twenty-two are how many times 11? How many times 2?

13. Twenty-four are how many times 12? How many times 2? How many times 8? How many times 3?

14. Twenty-six are how many times 2? How many times 13? What are the factors of 26?

15. Twenty-seven are how many times 9? How many times 3? What are the factors of 27?

16. Twenty-eight are how many times 14? How many times 2? What are the factors of 28?

17. Thirty are how many times 15? How many times 2? How many times 10? How many times 3? What are the factors of 30?

18. Thirty-three are how many times 11? How many times 3?

19. Thirty-four are how many times 2? How many times 17?

20. Thirty-six are how many times 18? How many times 12? How many times 9? How many times 6? How many times 4? How many times 3? How many times 2? What are the factors of 36?

21. Thirty-eight are how many times 19? How many times 2? What are the factors of 38?

22. Forty are how many times 20? How many times 10? How many times 8? How many times 5? How many times 4? How many times 2?

23. Forty-two are how many times 21? How many times 2? How many times 6? How many times 7? What are the factors of 42?

24. Forty-four are how many times 22? How many times 2? How many times 11? How many times 4? What are the factors of 44?

25. Forty-six are how many times 23? How many times 2? What are the factors of 46?

26. Forty-eight are how many times 24? How many times 16? How many times 12? How many times 8? How many times 6? How many times 4? How many times 3? How many times 2? What are the factors of 48?

27. Forty-nine are how many times 7? What are the factors of 49?

28. Fifty are how many times 25? How many times 10? How many times 5? What are the factors of 50?

29. Fifty-four are how many times 9? How many times 6? How many times 27? What are the factors of 54?

30. Fifty-six are how many times 7? How many times 8? How many times 28? What are the factors?

31. Sixty are how many times 10? How many times 12? How many times 5? What are the factors?

32. Sixty-four are how many times 8? How many times 16? How many times 4? What are the factors?

33. Seventy are how many times 7? How many times 10? What are the factors?

34. Seventy-two are how many times 8? How many times 9? What are the factors?

35. Eighty-four are how many times 12? How many times 7? What are the factors?

36. Ninety-six are how many times 12? How many times 8? What are the factors?

SECTION FOURTH.

LESSON I.

*In which Divisors are used from 2 to 5.**

1. How many 2's are there in 2. 2 is contained in 2, how many times?

2. How many 2's are there in 4? 2 in 4, how many times?

3. How many 2's are there in 6? 2 in 6, how many times?

4. How many 2's are there in 8? 2 in 8, how many times?

5. How many 2's are there in 10? 2 in 10, how many times?

6. How many 2's are there in 12? 2 in 12, how many times?

7. How many 3's are there in 3? 3 is contained in 3, how many times?

8. How many 3's are there in 6? 3 in 6, how many times?

9. How many 3's are there in 9? 3 in 9, how many?

10. How many 3's are there in 12? 3 in 12, how many?

11. How many 3's are there in 15? 3 in 15, how many times?

---

* Division is the process of finding how many times one number contains another.

The number to be divided is called the *dividend.*

The number by which we divide is called the *divisor.*

The number expressing how many times the dividend contains the divisor is called the quotient.

In case there is a number left, it is called the remainder. The divisor and quotient are *factors* of the *dividend.*

12. How many 3's are there in 18? 3 in 18, how many times?

13. How many 4's are there in 4? 4 is contained in 4, how many times?

14. How many 4's are there in 8? 4 in 8, how many times?

15. How many 4's are there in twelve? 4 in 12, how many times?

16. How many 4's are there in 16? 4 in 16, how many times?

17. How many 4's are there in 20? 4 in 20, how many times?

18. How many 4's are there in 24? 4 in 24, how many times?

19. How many 5's are there in 5? 5 is contained in 5, how many times?

20. How many 5's are there in 10? 5 in 10, how many times?

21. How many 5's are there in 15? 5 in 15, how many times?

22. How many 5's are there in 20? 5 in 20, how many times?

23. How many 5's are there in 25? 5 in 25, how many times?

24. How many 5's are there in 30? 5 in 30, how many times?

QUESTIONS.

1. *William has 8 apples, and divides them equally between two boys: how many does he give to each?

2. James has 12 peaches, and divides them equally between his two sisters: how many does he give to each?

* ANALYSIS.—Since 8 apples are to be divided equally between 2 boys, one will have as many apples as 2 is contained times in 8, which are 4: therefore, if 8 apples be equally divided between 2 boys, each boy will have 4 apples.

3. Charles has a basket containing 20 pears, and divides them equally between his father and mother: how many does he give to each?

4. A father bought 28 fish-hooks, and divided them between John and Charles: how many had each?

5. A mother has a dozen needles and gives an equal number to Jane and Mary: how many will each have?

6. A lady having two parlors, bought 24 chairs, and put an equal number in each room: how many were there in each room?

7. There are 16 boys in a school-room, and but 2 benches: how many boys must sit on each seat?

8. * How many peaches, at 2 cents apiece, can you buy for 18 cents?

9. At 3 cents apiece, how many oranges can you buy for 9 cents? How many can you buy for 12 cents? How many can you buy for 30 cents? How many can you buy for 24 cents?

10. A boy has 12 cents, and finds that he must give 3 cents apiece for tops: how many can he buy? If he has 21 cents, how many can he buy? If he has 24 cents, how many can he buy?

11. If 5 barrels of flour cost 30 dollars, how much will 1 barrel cost?†

* ANALYSIS.—Since one peach costs 2 cents, you can buy as many peaches for 18 cents as 2 is contained times in 18: 2 is contained in 18, 9 times; therefore, you can buy 9 peaches at 2 cents apiece for 18 cents.

† Since 5 barrels of flour cost 30 dollars, one barrel will cost as many dollars as 5 is contained times in 30: 5 is contained in 30, 6 times; therefore, if 5 barrels of flour cost 30 dollars, one barrel will cost 6 dollars.

NOTE.—The following rules result from the analysis of examples, 1, 8 and 11:

Ex. 1. *Divide the whole number of things by the number of parts into which they are to be divided; the quotient will be the number in each part.*

Ex. 8. *Divide the entire cost by the cost of a single thing; and the quotient is the number of things.*

Ex. 11. *Divided the entire cost by the number of things, and the quotient will be the cost of a single thing.*

12. If 4 yards of cloth cost 24 dollars, how much will 1 yard cost? How much will 2 yards cost? 3? 4? 6?

13. If 3 yards of ribbon cost 36 cents, how much will it cost a yard? How much will 11 yards cost?

14. If 4 pounds of beef cost 48 cents, how much will 1 pound cost? 2? 3? 5?

15. Twenty dollars are paid for 5 yards of cloth, how much is paid for 2 yards? 4? 6? 10?

16. Fifteen dollars are paid for 5 pairs of boots: how much must be paid for 1 pair?

17. If 4 apples be equally divided between 4 boys, how many will each have?

18. At 4 cents apiece, how many oranges can you buy for 8 cents? How many can you buy for 16 cents?

19. If it takes 4 sheets of paper for a book, how many books will 20 sheets make? How many will 28 sheets make? How many will 32 sheets make? How many will 36 sheets make? How many will 40 sheets make?

20. There are 4 benches in a school-room, and 20 scholars: how many must sit on each bench? If there be 24 scholars, how many must sit on each bench? If there are 32 scholars, how many must sit on each bench? If there be 36 scholars, how many must sit on each bench?

21. If John pays 4 cents for one top, how many tops will he buy for 12 cents? How many will he buy for 16 cents? How many will he buy for 20 cents? How many will he buy for 28 cents? How many will he buy for 40 cents?

22. If Charles gives 4 cents a quart for chestnuts, how many will he buy for 8 cents? How many for 16 cents? How many for 36 cents?

23. In a school-house there are 5 benches and 20

scholars: how many must sit on a bench? If there are 25 scholars, how many must sit on a bench? If there are 30, how many would there sit on a bench?

24. If cloth is 5 dollars a yard, how many yards can be purchased for 10 dollars? How much can be purchased for 20 dollars? For 30 dollars? For 40 dollars? For 50 dollars?

25. If flour is 5 dollars a barrel, how many barrels can be purchased for 15 dollars? How many barrels can be purchased for 20 dollars? How many barrels for 30 dollars?

26. If tape is 5 cents a bunch, how many bunches can be bought for 20 cents? How many bunches can be bought for 50 cents? How many bunches can be bought for 45 cents?

27. If 5 sheets of paper make a copy-book, how many books will 20 sheets make? How many books will 30 sheets make?

LESSON II.

In which Divisors are used from 6 *to* 10.

1. How many 6's are there in 6? 6 is contained in 6, how many times?

2. How many 6's are there in 12? 6 in 12, how many times?

3. How many 6's in 18? 6 in 18, how many times?

4. How many 6's in 24? 6 in 24, how many times?

5. How many 6's in 30? 6 in 30, how many times?

6. How many 6's in 36? 6 in 36, how many times?

7. How many 7's are there in 7? 7 is contained in 7, how many times?

8. How many 7's are there in 14? 7 in 14, how many times?

9. How many 7's are there in 21? 7 in 21, how many times?

10. How many 7's are there in 28? 7 in 28, how many times?

11. How many 7's are there in 35? 7 in 35, how many times?

12. How many 7's are there in 42? 7 in 42, how many times?

13. How many 8's are there in 8? 8 is contained in 8, how many times?

14. How many 8's are there in 16? 8 in 16, how many times?

15. How many 8's are there in 24? 8 in 24, how many times?

16. How many 8's are there in 32? 8 in 32, how many times?

17. How many 8's are there in 40? 8 in 40, how many times?

18. How many 8's are there in 48? 8 in 48, how many times?

19. How many 9's are there in 9? 9 in 9, how many times?

20. How many 9's are there in 18? 9 in 18, how many times?

21. How many 9's are there in 27? 9 in 27, how many times?

22. How many 9's are there in 36? 9 in 36, how many times?

23. How many 9's are there in 45? 9 in 45, how many times?

24. How many 9's are there in 54? 9 in 54, how many times?

25. How many 10's are there in 30? 10 in 30, how many times?

26. How many 10's are there in 60? 10 in 60, how many times?

27. How many 10's are there in 100? 10 in 100, how many times?

QUESTIONS.

1. If 6 sheets of paper make a copy-book, how many books will 12 sheets make? How many books will 24 sheets make? How many books will 30 sheets make? How many books will 48 sheets make? How many books will 60 sheets make?

2. If 1 yard of broadcloth costs 6 dollars, how many yards can be bought for 30 dollars? How many yards can be bought for 36 dollars? How many yards for 42 dollars? How many yards for 54 dollars? How many for 60 dollars?

3. If a man travels 6 miles in 1 hour, how many hours will it take him to travel 12 miles? How many hours will it take him to travel 24 miles? How long will it take him to travel 30 miles? How long to travel 54 miles? How long to travel 60 miles?

4. Forty-two apples are divided equally among 6 boys: how many does each one receive?

5. If 54 peaches be divided equally between 6 boys, how many will each receive?

6. If a yard of ribbon costs 6 cents, how many yards can be bought for 24 cents? How many for 30 cents?

7. If you have 28 dollars, how many yards of cloth can you buy at 7 dollars a yard? How many yards for 35 dollars? How many for 70 dollars?

8. If you have 16 apples to divide among 8 boys, how many do you give to each? If you have 32,

how many? If you have 64, how many? If you have 96, how many?

9. A laborer engaged to work for 12 dollars a month; at the end of the time he received 96 dollars, how many months did he work?

10. James engaged to work for 9 cents a day, and at the end of the time received 72 cents: how many days did he work?

11. If a man can do a piece of work in 56 hours, how many days will it take him to do it, working 7 hours a day?

12. If two boys are 36 yards apart, and the one behind gains on the other 4 yard a minute, how many minutes before he will overtake him?

13. A man has 84 pounds of butter to be put in seven jars: how much must be put in each jar?

14. A lady paid 108 cents for 9 yards of ribbon: how much did she pay a yard?

15. If a milliner pays 121 cents for 11 yards of ribbon: how much does she pay for 1 yard?

16. How many dresses can be cut from 132 yards of silk, if each dress contains 12 yards?

LESSON III.

Addition, Subtraction, Multiplication, Division.

1. Two in 5, how many times, and what over?
2. Two in 7, how many times, and what over?*
3. Four in 15, how many times?
4. Five in 19, how many times?
5. Five in 36, how many times?
6. Seven in 42, how many times?
7. Nine in 60 how many times?

---

* When there is a remainder, let it be simply mentioned.

8. Seven in 64, how many times?
9. Ten in 55, how many times?
10. Six in 70, how many times?
11. Nine in 100, how many times?
12 Five in 56, how many times?
13. Twelve are how many times 2?
14. Eighteen are how many times 6?
15. Nineteen plus 6, are how many times 5?
16. Twenty plus 8, are how many times 7? How many times 4?
17. Thirty less 6, are how many times 6? How many times 4? How many times 3?
18. Sixty less 5, are how many times 11?
19. Ninety plus 9, are how many times 11?
20. Eighty-seven plus 3, are how many times 10?
21. Forty-five plus 4, are how many times 7?
22. Sixty-nine plus 15, are how many times 12? How many times 6? How many times 7?
23. Six times 7 less 2, are how many times 10?
24. Forty plus 4 times 6, are how many times 8?
25. Fifty plus 3 times 4, plus 2, are how many times 8?
26. Five times 6 plus 4 times 9, are how many times 1?
27. Seven times 8 plus 4, are how many times 6? 12? 10? 5?
28. Eight times 5 plus 5, are how many times 9? 5?
29. Five times 11 less 5, are how many times 2? 10? 5?
30. Forty-six less 3 times 2, are how many times 10? 8? 5? 4? 2?
31. Seven times 9 plus 3 times 4, are how many times 25?
32. Six times 4 less 3 times 4, are how many times 3? 4? 6?

QUESTIONS.

1. If 2 yards of cloth cost 6 dollars, what will 8 yards cost?*

2. If 3 oranges cost 12 cents, what will 11 cost?

3. If 4 boxes of raisins can be bought for 16 dollars, how much will 9 boxes cost?

4. If 7 pounds of sugar cost 56 cents, what will 12 pounds cost?

5. What will 6 lemons cost if 8 cost 24 cents?

6. James has apples worth 2 cents apiece, how many must he give for 6 oranges worth 3 cents apiece?

7. How many eggs at 8 cents a dozen must be given for 12 pounds of sugar worth 6 cents a pound?

8. How many knives at 2 shillings apiece are worth 4 axes at 8 shillings apiece?

9. How much barley at 5 shillings a bushel must be given for 6 bushels of wheat at 10 shillings a bushel?

10. A farmer bought 4 yards of cloth at 3 dollars a yard and paid in labor at 2 dollars a day; how many days must he labor?

11. John had oranges worth 4 cents a piece which he gave to James for 2 quarts of cherries worth 8 cents a quart; how many oranges did he give for the cherries?

NOTE.—If any number be divided into two equal parts, one of the parts is called one half. If it be divided into three equal parts, one of the parts is called *one third*, and two of the parts are called *two thirds*. If it be divided into four equal parts, one of the parts is called *one fourth*, two of the parts, *two fourths*, three of them, *three fourths*, &c., &c.

* ANALYSIS.—One yard will cost one half as much as two yards. If two yards cost 6 dollars, one yard will cost one-half of six dollars, which are 3 dollars; 8 yards will cost 8 times as much as 1 yard: if 1 yard costs 3 dollars, 8 yards will cost 8 times 3 dollars which are 24 dollars: therefore, if 2 yards of cloth cost 6 dollars, 8 yards will cost 24 dollars.

12. How many yards of cotton cloth, at 8 cents a yard, must be given for 6 pounds of butter worth 12 cents a pound?

13. A farmer bought 4 yards of broadcloth worth 5 dollars a yard, and paid for them in calves worth 4 dollars apiece. How many must he give?

14. How many barrels of flour worth 6 dollars a barrel must be given for 8 barrels of fish worth 3 dollars a barrel?

15. A man bought 18 oranges at the rate of 3 for 5 cents. How much did they come to?*

16. How many pears at the rate of 2 for 3 cents can be bought for 24 cents?

17. James bought 20 marbles at the rate of 5 for 6 cents. How much did they cost?

18. How much cloth worth 2 dollars a yard, must be given for 2 firkins of butter worth 18 dollars a firkin?

19. What will 8 pounds of beef cost, if 6 pounds cost 54 cents?

20. A farmer has 6 dozen of eggs worth 12 cents a dozen, and wishes to exchange them for nutmegs at 3 cents apiece: how many shall he receive?

21. A man bought 4 barrels of flour at 7 dollars a barrel, and wishes to exchange them for cloth at 2 dollars a yard: how many yards of cloth should he receive?

22. Four men bought a boat for 60 dollars: they paid 12 dollars for repairing her, and then sold her so that they made 5 dollars: what did they get for her? How much more than they gave?

---

* ANALYSIS.—Since 3 oranges cost 5 cents, 18 oranges will cost as many times 5 cents as 3 is contained times in 18, which is 6 times: therefore, 18 oranges will cost 6 times 5 cents which are 30 cents.

23. A laborer engaged to work for a year at 144 dollars a month, but was sent away at the end of 7 months: how much should he receive? How much less than if he had staid the entire year?

24. A drover has 8 calves for which he paid 40 dollars, and 9 sheep for which he paid 27 dollars: what did he pay apiece for the calves? What for the sheep? How much for all?

25. If 8 firkins of butter are worth 96 dollars, how many firkins must be given for 3 barrels of sugar worth 20 dollars a barrel?

26. Six men agree to do a piece of work for 90 dollars: it turns out that each man makes 5 dollars by the bargain: what was the cost of doing the work?

27. A man has 40 eggs, from which he wishes to raise chickens: his eggs are worth a cent apiece, and his chickens, when hatched, will be worth 3 cents apiece. Now, if 15 of his eggs prove addle, how much will he gain?

28. How much honey at 16 cents a pound must be given for 6 pounds of coffee at 8 cents a pound?

29. How many hats at 4 dollars apiece, can a man buy if he gives in payment seven handkerchiefs at 1 dollar apiece, and 5 pair of boots at 6 dollars a pair? What will he have left?

30. A grocer purchased 4 barrels of sugar at 12 dollars a barrel, and 5 hogsheads of molasses at 20 dollars a hogshead, and sold the whole for 150 dollars: did he make or lose, and how much?

31. Five men agree to do a piece of work for 60 dollars, each to receive an equal part. When the [illegible] is half done, two of the men quit, and the other finish it: how much should each receive?

[illegible] Three men made up a purse. The first put in [illegible] dollars; the second twice as much as the first, and the third twice as much as the second: how much was put in in all? What would each have put in had they contributed equally?

33. Mr. Johnson bought 14 yards of broadcloth at 4 dollars a yard, and having cut off 9 yards, sold the remainder for one half of what he paid for the whole less 3 dollars: how much more was this for each yard than he gave?

34. If 6 men can do a piece of work in 9 days, how long will it take one man to do it?*

35. How many men in 9 days can perform as much labor as 12 men can in 6 days?

36. If a barrel of flour will last 5 men 25 days, how long will it last 9 men?

37. A man has a piece of work which would employ 3 men for 8 days: how many men can do it in 1 day? In 4 days?†

38. If 8 men can do a piece of work in 5 days, how long will it take one man to do it? How many men will do it in one day?

39. A piece of work requires 9 men for 10 days, how many men can do the same work in one day? In 3 days? In 6 days?

40. If 6 men can build a wall 12 rods long in 8 days, how long will it take one man to build it? How many men can build it in 1 day? In 3 days? In 12 days?

---

* ANALYSIS.—It will take 1 man 6 times as long to do the same work as it will 6 men: if it takes 6 men 9 days, it will take 1 man 6 times 9 days, which are 54 days: therefore, if it take 6 men 9 days to do a piece of work, it will take 1 man 54 days to do the same work.

† ANALYSIS.—It will take 8 times as many men to work in 1 day as it will to do it in 8 days: if it takes 3 days, it will take 8 times 3 men, which are 24 men, work in one day: therefore, if 3 men can do a piece of 8 days, 24 men can do the same work in 1 day.

LESSON IV.

Combinations.

The classes are to be drilled in the following tables until they are fully understood. The tables are to be read across the page: thus, 1 and 1 are two; 1 from 2 leaves one; once 1 is one; 1 in 1, once; and similarly for the other lines.*

1 and 1;	1 from 2;	once 1;	1 in 1
1 and 2;	1 from 3;	once 2;	1 in 2
1 and 3;	1 from 4;	once 3;	1 in 3
1 and 4;	1 from 5;	once 4;	1 in 4
1 and 5;	1 from 6;	once 5;	1 in 5
1 and 6;	1 from 7;	once 6;	1 in 6
1 and 7;	1 from 8;	once 7;	1 in 7
1 and 8;	1 from 9;	once 8;	1 in 8
1 and 9;	1 from 10;	once 9;	1 in 9
1 and 10;	1 from 11;	once 10;	1 in 10
2 and 1;	2 from 3;	two times 1;	2 in 2
2 and 2;	2 from 4;	two times 2;	2 in 4
2 and 3;	2 from 5;	two times 3;	2 in 6
2 and 4;	2 from 6;	two times 4;	2 in 8
2 and 5;	2 from 7;	two times 5;	2 in 10
2 and 6;	2 from 8;	two times 6;	2 in 12
2 and 7;	2 from 9;	two times 7;	2 in 14
2 and 8;	2 from 10;	two times 8;	2 in 16
2 and 9;	2 from 11;	two times 9;	2 in 18
2 and 10;	2 from 12;	two times 10;	2 in 20

* SUGGESTION.—The object in placing these tables here is to drill the pupil in changing his mind from one subject to another. In each line he begins with addition, passes to subtraction, then multiplication, and then to division. The difficulty of doing [illegible] *accurately*, and rapidly, will illustrate fully the value of the [illegible] mind must be well disciplined that can go through [illegible]les without a mistake.

3 and 1 ;	3 from 4 ;	3 times 1 ;	3 in 3
3 and 2 ;	3 from 5 ;	3 times 2 ;	3 in 6
3 and 3 ;	3 from 6 ;	3 times 3 ;	3 in 9
3 and 4 ;	3 from 7 ;	3 times 4 ;	3 in 12
3 and 5 ;	3 from 8 ;	3 times 5 ;	3 in 15
3 and 6 ;	3 from 9 ;	3 times 6 ;	3 in 18
3 and 7 ;	3 from 10 ;	3 times 7 ;	3 in 21
3 and 8 ;	3 from 11 ;	3 times 8 ;	3 in 24
3 and 9 ;	3 from 12 ;	3 times 9 ;	3 in 27
3 and 10 ;	3 from 13 ;	3 times 10 ;	3 in 30
4 and 1 ;	4 from 5 ;	4 times 1 ;	4 in 4
4 and 2 ;	4 from 6 ;	4 times 2 ;	4 in 8
4 and 3 ;	4 from 7 ;	4 times 3 ;	4 in 12
4 and 4 ;	4 from 8 ;	4 times 4 ;	4 in 16
4 and 5 ;	4 from 9 ;	4 times 5 ;	4 in 20
4 and 6 ;	4 from 10 ;	4 times 6 ;	4 in 24
4 and 7 ;	4 from 11 ;	4 times 7 ;	4 in 28
4 and 8 ;	4 from 12 ;	4 times 8 ;	4 in 32
4 and 9 ;	4 from 13 ;	4 times 9 ;	4 in 36
4 and 10 ;	4 from 14 ;	4 times 10 ;	4 in 40
5 and 1 ;	5 from 6 ;	5 times 1 ;	5 in 5
5 and 2 ;	5 from 7 ;	5 times 2 ;	5 in 10
5 and 3 ;	5 from 8 ;	5 times 3 ;	5 in 15
5 and 4 ;	5 from 9 ;	5 times 4 ;	5 in 20
5 and 5 ;	5 from 10 ;	5 times 5 ;	5 in 25
5 and 6 ;	5 from 11 ;	5 times 6 ;	5 in 30
5 and 7 ;	5 from 12 ;	5 times 7 ;	5 in 35
5 and 8 ;	5 from 13 ;	5 times 8 ;	5 in 40
5 and 9 ;	5 from 14 ;	5 times 9 ;	5 in 4[illegible]
5 and 10 ;	5 from 15 ;	5 times 10 ;	5 in 5[illegible]
6 and 1 ;	6 from 7 ;	6 times 1 ;	6 in [illegible]
6 and 2 ;	6 from 8 ;	6 times 2 ;	6 in [illegible]
6 and 3 ;	6 from 9 ;	6 times 3 ;	6 in [illegible]
6 and 4 ;	6 from 10 ;	6 times [illegible]	
6 and 5 ;	6 from 11 ;	6 times [illegible]	

6 and 6 ; 6 from 12 ; 6 times 6 , 6 in 36
6 and 7 ; 6 from 13 ; 6 times 7 ; 6 in 42
6 and 8 ; 6 from 14 ; 6 times 8 ; 6 in 48
6 and 9 ; 6 from 15 ; 6 times 9 ; 6 in 54
6 and 10 ; 6 from 15 ; 6 times 10 ; 6 in 60

7 and 1 ; 7 from 8 ; 7 times 1 ; 7 in 7
7 and 2 ; 7 from 9 ; 7 times 2 ; 7 in 14
7 and 3 ; 7 from 10 ; 7 times 3 ; 7 in 21
7 and 4 ; 7 from 11 ; 7 times 4 : 7 in 28
7 and 5 ; 7 from 12 ; 7 times 5 ; 7 in 35
7 and 6 ; 7 from 13 ; 7 times 6 ; 7 in 42
7 and 7 ; 7 from 14 ; 7 times 7 ; 7 in 49
7 and 8 ; 7 from 15 ; 7 times 8 ; 7 in 56
7 and 9 ; 7 from 16 ; 7 times 9 ; 7 in 63
7 and 10 ; 7 from 17 ; 7 times 10 ; 7 in 70

8 and 1 ; 8 from 9 ; 8 times 1 ; 8 in 8
8 and 2 ; 8 from 10 ; 8 times 2 ; 8 in 16
8 and 3 ; 8 from 11 ; 8 times 3 ; 8 in 24
8 and 4 ; 8 from 12 ; 8 times 4 ; 8 in 32
8 and 5 ; 8 from 13 ; 8 times 5 ; 8 in 40
8 and 6 ; 8 from 14 ; 8 times 6 ; 8 in 48
8 and 7 ; 8 from 15 ; 8 times 7 ; 8 in 56
8 and 8 ; 8 from 16 ; 8 times 8 ; 8 in 64
8 and 9 ; 8 from 17 ; 8 times 9 ; 8 in 72
8 and 10 , 8 from 18 ; 8 times 10 ; 8 in 80

9 and 1 ; 9 from 10 ; 9 times 1 ; 9 in 9
9 and 2 ; 9 from 11 ; 9 times 2 ; 9
9 and 3 ; 9 from 12 ; 9 times 3 ;
9 and 4 ; 9 from 13 ; 9 times 4 ;
9 and 5 ; 9 from 14 ; 9 times 5 ;
9 and 6 ; 9 from 15 ; 9 times 6 ; 9
9 and 7 ; 9 from 16 ; 9 times 7 ;
8 ; 9 from 17 ; 9 times
9 fro

10 and 1 ; 10 from 11 ; 10 times 1 ; 10 in 10
10 and 2 ; 10 from 12 ; 10 times 2 ; 10 in 20
10 and 3 ; 10 from 13 ; 10 times 3 ; 10 in 30
10 and 4 ; 10 from 14 ; 10 times 4 ; 10 in 40
10 and 5 ; 10 from 15 ; 10 times 5 ; 10 in 50
10 and 6 ; 10 from 16 ; 10 times 6 ; 10 in 60
10 and 7 ; 10 from 17 ; 10 times 7 ; 10 in 70
10 and 8 ; 10 from 18 ; 10 times 8 ; 10 in 80
10 and 9 ; 10 from 19 ; 10 times 9 ; 10 in 90
10 and 10 ; 10 from 20 ; 10 times 10 ; 10 in 100

11 and 1 ; 11 from 12 ; 11 times 1 ; 11 in 11
11 and 2 ; 11 from 13 ; 11 times 2 ; 11 in 22
11 and 3 ; 11 from 14 ; 11 times 3 ; 11 in 33
11 and 4 ; 11 from 15 ; 11 times 4 ; 11 in 44
11 and 5 ; 11 from 16 ; 11 times 5 ; 11 in 55
11 and 6 ; 11 from 17 ; 11 times 6 ; 11 in 66
11 and 7 ; 11 from 18 ; 11 times 7 ; 11 in 77
11 and 8 ; 11 from 19 ; 11 times 8 ; 11 in 88
11 and 9 ; 11 from 20 ; 11 times 9 ; 11 in 99
11 and 10 ; 11 from 21 ; 11 times 10 ; 11 in 110
11 and 11 ; 11 from 22 ; 11 times 11 ; 11 in 121
11 and 12 ; 11 from 23 ; 11 times 12 ; 11 in 132

12 and 1 ; 12 from 13 ; 12 times 1 ; 12 in 12
12 and 2 ; 12 from 14 ; 12 times 2 ; 12 in 24
12 and 3 ; 12 from 15 ; 12 times 3 ; 12 in 36
12 and 4 ; 12 from 16 ; 12 times 4 ; 12 in 48
12 and 5 ; 12 from 17 ; 12 times 5 ; 12 in 60
6 ; 12 from 18 ; 12 times 6 ; 12 in 72
7 ; 12 from 19 ; 12 times 7 ; 12 in 84
8 ; 12 from 20 ; 12 times 8 ; 12 in 96
9 ; 12 from 21 ; 12 times 9 ; 12 in 108
10 ; 12 from 22 ; 12 times 10 ; 12 in 12[illegible]
11 · 12 from 23 ; 12 times 11 ; 12 in 132
12 times 12

SECTION FIFTH.

LESSON I.

UNITED STATES MONEY.

10 mills make . . .	1 cent . . .	*ct.*
10 cents	1 dime . . .	*d.*
10 dimes	1 dollar . .	$.
10 dollars	1 eagle . . .	*E.*

1. How many mills are there in 2 cents? In 8 cents? In half a cent? In five cents?

2. How many cents are there in 10 mills? In 15 mills? In 65 mills?

3. How many cents are there in 5 dimes? In 6 dimes? In 8 dimes? In 10 dimes?

4. How many dimes are there in 10 cents? In 12 cents? In 16 cents? In 20 cents?

5. How many dimes in 1 dollar? In 2 dollars? In 3 dollars? In 4 dollars? In 5 dollars?

6. How many dollars in 1 eagle? In 2 eagles? In 5 eagles? In 6 eagles? In 9 eagles?

7. How many eagles in 20 dollars? In 30 dollars? In 50 dollars? In 60 dollars?

8. If 3 yards of cloth cost 24 dimes, what will 7 yards cost?

9. If 7 pounds of tea cost 42 dimes, what will 9 pounds cost?

10. If 6 cows cost eighteen eagles, how much will they cost apiece? What will 9 cost?

ENGLISH STERLING MONEY.

4 farthings make . .	1 penny . .	*d.*
12 pence	1 shilling . .	*s.*
20 shillings	1 pound . .	£.
21 shillings	1 guinea . .	

6

1. How many farthings are there in 1 penny? In 2 pence? In 4? In 6? In 8?

2. How many pence in 4 farthings? In 8 farthings? In 12? In 14? In 16?

3. How many pence are there in 1 shilling? In 2 shillings? In 3 shillings? In 4 shillings?

4. How many shillings are there in 12 pence? In 18? In 20? In 24? In 26 pence?

5. How many shillings are there in 1 pound? In 2 pounds? In 3? In 5? In 4 pounds?

6. How many pounds are there in 20 shillings? In 40 shillings? In 60 shillings? In 80 shillings?

7. How many guineas in 21 shillings? In 42 shillings? In 84 shillings? In 63 shillings?

LESSON II.

TABLE OF TROY WEIGHT.

24 grains, *gr.* make .	1 pennyweight,	.	*pwt.*
20 pennyweights . .	1 ounce,	. . .	*oz.*
12 ounces	1 pound,	. . .	*lb.*

1. How many grains are there in 1 pennyweight? In 2 pennyweights? In 3 pennyweights? In 4 pennyweights?

2. How many pennyweights are there in 24 grains? In 48 grains? In 72 grains? In 96 grains?

3. How many pennyweights in 1 ounce? In 2 ounces? In 3? In 4? In 5?

4. How many ounces are there in 20 pennyweights? In 40 pennyweights? In 60? In 80 pennyweights?

5. How many ounces are there in 1 pound? In 2 pounds? In 3 pounds? In 4 pounds? In 5 pounds?

6. How many pounds in 12 ounces? In 24? In 36? In 48? In 60?

7. How many pennyweights in 4 ounces? In 2 ounces? How many grains in 1 ounce? In 2 ounces? In 3 ounces?

TABLE OF APOTHECARIES' WEIGHT.

20 grains, *gr.* make .	1 scruple, . . .	℈.
3 scruples	1 dram, . . .	ʒ.
8 drams	1 ounce, . . .	℥.
12 ounces	1 pound, . . .	℔.

1. How many grains are there in 1 scruple? In 2 scruples? In 3? In 4?

2. How many scruples in 20 grains? In 40? In 60? In 80?

3. How many scruples in 1 dram? In 2? In 3? In 4, how many?

TABLE OF AVOIRDUPOIS WEIGHT.

16 drams, *dr.* make .	1 ounce, . . .	*oz.*
16 ounces	1 pound, . . .	*lb.*
25 pounds	1 quarter, . . .	*qr.*
4 quarters	1 hundred weight,	*cwt.*
20 hundred weight .	1 ton, ...	*T.*

1. How many drams in an ounce? How many ounces in a pound? How many pounds in a quarter? How many quarters in a hundred? How many hundred in a ton?

2. How many drams in 2 ounces? How many ounces in 2 pounds? How many pounds in 2 quarters? How many quarters in 2 hundred? How many hundreds in two tons?

3. How many ounces in 6 pounds? How many in 4 pounds? How many tons in 60 hundred? How many quarters in 5 tons? How many pounds in 3 hundred and 3 quarters? How many ounces in 11 pounds?

LESSON III.

TABLE OF LONG MEASURE.

12 inches, *in.* make .	1 foot, ...	*ft.*
3 feet	1 yard, ...	*yd.*
$5\frac{1}{2}$ yards, or $16\frac{1}{2}$ feet .	1 rod, ...	*rd.*
40 rods	1 furlong, . . .	*fur.*
8 furlongs, or 320 rods	1 mile, ...	*m.*
3 miles	1 league, . . .	*L.*
60 geographical or $69\frac{1}{2}$ statute miles . .	1 degree, .	*deg.* or °.
360 degrees	a great circle.	

1. How many feet in 24 inches? In 6 yards how many? How many yards in 15 feet? In 24? How many rods in 2 furlongs? How many furlongs in 160 rods? How many miles in 9 leagues? How many leagues in 54 miles?

TABLE OF SQUARE MEASURE.

144 square inches, *sq. in.* make	1 square foot,	*sq. ft.*
9 square feet	1 square yard,	*sq. yd.*
$30\frac{1}{4}$ square yards	1 square pole, .	*P.*
40 square rods	1 rood, . . .	*R.*
4 roods	1 acre, . . .	*A.*
640 acres	1 square mile, .	*M.*

1. How many square inches in 2 square feet?
2. How many square feet in 2 square yards? In 8? In 4? In 5? In 9? In 10?
3. How many square rods in 2 roods? How many in 3 roods? In 5?
4. How many roods in 80 square rods? In 120? In 160, how many?
5. How many roods in 2 acres? How many in 3 acres? How many in 5 acres?
6. How many acres in 8 roods? In 12, how many? In 20?
7. How many acres make 4 square miles?

TABLE OF CLOTH MEASURE.

$2\frac{1}{4}$	inches, *in.*	make	1 nail, ...	*na.*
4	nails		1 quarter of a yard,	*qr.*
4	quarters . . .		1 yard, ...	*yd.*
3	quarters . . .		1 Ell Flemish,	*E. Fl.*
5	quarters . .		1 Ell English, .	*E. E.*
6	quarters . . .		1 Ell French, .	*E. Fr.*

1. How many inches make 2 nails? 3 nails? 4 nails?

2. How many nails make a quarter of a yard? How many make 2 quarters? How many make 3 quarters?

3. How many quarters make 2 yards? 3 yards? 4 yards? 5 yards?

4. How many quarters make an Ell Flemish? How many make 2 Ells? 3 Ells? 4? 5? 6? 7?

5. How many Ells Flemish in 3 quarters of a yard? In 6 quarters? In 9? In 12? In 15?

TABLE OF WINE MEASURE.

4	gills, *gi.*	make	1 pint, . .	*pt.*
2	pints		1 quart, . . .	*qt.*
4	quarts		1 gallon, . .	*gal.*
$31\frac{1}{2}$	gallons		1 barrel, . .	*bar.*
63	gallons		1 hogshead, .	*hhd.*
2	hogsheads		1 pipe, . . .	*pi.*
2	pipes, or 4 hogsheads		1 tun, . . .	*tun.*

1. How many gills make a pint? How many make 2 pints? 3? 4? 5? 6?

2. How many pints in a quart? In 2 quarts? In 3? In 4? In 6? In 7?

3. How many quarts in 2 pints? In 4? In 6? In 8? In 10? In 12?

4. How many quarts in 1 gallon? In 2? In 3? In 4? In 5? In 7?

5. How many gallons in 4 quarts? In 8? In 12? In 16? In 20?

6. How many gallons in a barrel?

7. How many gallons in a hogshead? In 2?

8. How many hogsheads in a pipe? In 2 pipes? In 3? In 4? In 5?

9. How many pipes in 2 hogsheads? In 4?

10. How many pipes in 1 tun? In 2 tuns? In 3?

TABLE OF ALE AND BEER MEASURE.

2 pints, *pt.*	make	1 quart, ...	*qt.*
4 quarts		1 gallon, ...	*gal.*
36 gallons		1 barrel, ...	*bar.*
54 gallons		1 hogshead, . .	*hhd.*

1. How many pints in 2 quarts? In 3? In 4? In 5?

2. How many quarts in 2 pints? In 4? In 6? In 8? In 10? In 12?

3. How many quarts in 2 gallons? In 4? In 5?

4. How many gallons in 4 quarts? In 8? In 12?

TABLE OF DRY MEASURE.

2 pints *pt.*	make	1 quart, ...	*qt.*
8 quarts		1 peck, ...	*pk.*
4 pecks		1 bushel, . . .	*bu.*
36 bushels		1 chaldron, . . .	*ch.*

1. How many pints in 2 quarts? In 3? In 4? In 5? In 6?

2. How many quarts in 4 pints? In 6? In 8? In 10?

3. How many quarts in 2 pecks? In 3? In 4? In 5?

4. How many pecks in 16 quarts? In 24? In 32? In 40?

TIME TABLE.

60 seconds, *sec.*	make	1 minute,	. . .	*m.*
60 minutes	. . .	1 hour,	. . . .	*hr.*
24 hours	. . . . .	1 day,	. . . .	*da.*
7 days	. . . . .	1 week,	. . . .	*wk.*
4 weeks	. . . . .	1 month,	. . .	*mo.*
52 weeks	. . . . .	1 year,	. . . .	*yr.*
100 years	. . . . .	1 century,	. . .	*C.*

The year is also divided into twelve calendar months, which contain an unequal number of days.

Winter,	1 month	January, . . .	31
	2 . . .	February, . . .	28*
Spring,	3 . . .	March, ...	31
	4 . . .	April, ...	30
	5 . . .	May, ...	31
Summer,	6 . . .	June, ...	30
	7 . . .	July	31
	8 . . .	August, . . .	31
Autumn,	9 . . .	September, . .	30
	10 . . .	October, . . .	31
	11 . . .	November, . .	30
Winter,	12 . . .	December, . .	31
		Total	365

Thirty days hath September,
April, June, and November;
All the rest have thirty-one,
Excepting February, twenty-eight alone.

* February has 29 days in leap-year

SECTION SIXTH.

LESSON I.

Of the Fraction One Half.

1. If an apple be divided into two equal parts, one of the parts is called one half. What are the 2 parts called?

2. How many halves are there in one apple?

3. If a pear be divided into two equal parts, what are the parts called?

4. How many halves are there in a pear?

5. How many halves are there in 1? In 1 and a half, how many?

6. How many halves are there in 2 pears? In 2 and a half, how many?*

7. How many halves are there in 3 things? In 3 and a half, how many?

8. How many halves are there in 4? In 4 and a half, how many?

9. How many halves are there in 5? In 5 and a half, how many?

10. How many halves are there in 6? In 6 and a half, how many?

11. How many halves are there in 7? In 7 and a half, how many?

12. How many halves in 8? In 8 and a half?

13. How many halves in 9? In 9 and a half?

14. How many halves in 10? In 10 and a half?

15. How many halves in 11 and a half? In 12 and a half?

16. How many whole things are there in 2 halves? In 4 halves?† In 6 halves? In 12 halves? In 18 halves? In 24? In 30? In 28? In 17? In 9? In 13? In 15?

* ANALYSIS.—There are 2 halves in 1; therefore, in any number of things there are twice as many halves as whole things. In 2 and a half there are 5 halves.

† ANALYSIS.—Since there is 1 whole thing in two halves, the number of whole things are found by dividing the halves by 2.

QUESTIONS.

1. If one half an orange cost 1 cent, how much will 3 halves cost?

2. If one half a pine-apple cost 4 cents, how much will 1 pine-apple cost?

3. If one half a yard of cloth cost 2 dollars, how much will 6 yards cost? 4 yards? 5 yards?

4. If one and one half yards of cloth cost 6 dollars, what will 5 yards cost?*

5. James gave 3 and a half cents for a top, 6 and a half cents for paper, 9 half cents for a pencil, and 7 half cents for an orange: how much did he pay in all?

6. Ann bought a skein of silk for 9 half cents, a spool of thread for 7 and a half cents, a thimble for 15 half-cents: how much did she pay in all?

LESSON II.

Of the Fraction One Third.

1. If an apple be divided into three equal parts, one of the parts is called one third: what are 2 of the parts called? 3 of them?

2. How many thirds are there in one apple?

3. How many thirds are there in 1?

4. How many thirds are there in 2? In 2 and one third? Why?

5. How many thirds are there in 3? In 3 and two thirds? In 3 and one third? Why?

---

* Analysis.—One and one half yards are equal to 3 half yards. One half yard will cost one third as much as 3 half yards. If 3 half yards cost 6 dollars, one half yard will cost one third of 6 dollars, which are 2 dollars. If one half a yard costs 2 dollars, one yard will cost 2 times 2 dollars, which are 4 dollars; and if 1 yard costs 4 dollars, 5 yards will cost 5 times 4 dollars, which are 20 dollars.

6. How many thirds are there in 4? In 4 and one third? In 4 and two thirds?

7. How many thirds are there in 5? In 5 and two thirds? In 5 and one third?

8. How many thirds are there in 6? In 6 and one third? In 6 and two thirds?

9. How many thirds are there in 7? In 7 and one third? In 7 and two thirds?

10. How many thirds are there in 8? In 8 and one third? In 8 and two thirds?

11. How many thirds are there in 9? In 9 and one third? In 9 and two thirds?

12. How many thirds are there in 10? In 10 and two thirds?

13. How many thirds are there in 11? In 11 and two thirds?

14. How many thirds are there in 12? In 12 and one third? In 12 and two thirds?

15. How many whole things are there in 6 thirds? In 9 thirds? In 15 thirds? In 18 thirds? In 8 thirds? In 7 thirds?

16. How many whole things in 14 thirds? In 16 thirds? In 25 thirds? In 17 thirds? In 19 thirds?

17. How many thirds in 9 and 6 thirds? In 7 and 5 thirds? In 6 and 11 thirds?

18. How many thirds in 6 thirds and 5 thirds? In 9 thirds, and 4 and 2 thirds, how many? In 16 thirds and 9 thirds and 5 thirds, how many thirds?

QUESTIONS.

1. If one third of an orange cost one cent, what will the whole orange cost? What will two oranges cost?

2. If one third of a yard of cloth cost two dollars, what will three yards cost?

3. If one third of a barrel of flour cost two dollars, what will six barrels cost?

4. If two thirds of a pound of tea cost 40 cents, what will 2 pounds cost?

5. If 3 and two thirds pounds of sugar cost 33 cents, how much will 8 pounds cost?

If 3 and one third pounds of coffee cost 40 cents, how much will 9 pounds cost?

7. If 8 and two thirds yards of cloth cost 52 dollars, how much will 7 yards cost?

8. John gave two and two thirds dimes for a penknife: how much did he pay for 3 penknives?

9. If 2 sheep cost 5 and one third dollars, what will 6 cost?

10. If 1 pound of honey cost 12 and two third cents, what will 9 pounds cost?

LESSON III.

Of the Fraction One Fourth.

1. If an apple be divided into four equal parts, one of the parts is called one fourth. What are 2 of them called? 3 of them? 4 of them?

2. How many fourths are there in 1?

3. How many fourths are there in 1 and one fourth?

4. How many fourths are there in 2? In 2 and one fourth? In 2 and three fourths?

5. How many fourths are there in 3? In 3 and two fourths? In 3 and three fourths?

6. How many fourths are there in 4? In 4 and one fourth? In 4 and two fourths?

7. How many fourths are there in 5? In 5 and one fourth? In 5 and three fourths?

8. How many fourths are there in 6? In 6 and one fourth? In 6 and three fourths?

9. How many fourths are there in 7? In 7 and one fourth. In 7 and three fourths?

10. How many fourths are there in 8? In 8 and one fourth? In 8 and two fourths? In 8 and three fourths?

11. How many fourths are there in 9? In 9 and one fourth? In 9 and two fourths? In 9 and three fourths?

12. How many fourths are there in 10? In 10 and one fourth? In 10 and two fourths? In 10 and three fourths?

13. How many whole things in 4 fourths? In 8 fourths? In 20 fourths? In 17 fourths? In 19 fourths?

14. How many whole things in 12 fourths? In 25 fourths? In 30 fourths? In 35 fourths? In 48 fourths? In 60 fourths? In 56 fourths?

QUESTIONS.

1. If one fourth of a yard of cloth costs 2 dollars, how much will 1 yard cost? How much will 4 yards cost?

2. If 3 fourths of a pound of coffee cost 12 cents, what will 5 pounds cost?

3. If 7 fourths yards of cloth cost 14 dollars, what will 9 yards cost?

4. If 3 yards of cloth cost 3 and 3 fourth dollars, what will 5 yards cost?

5. How many fourths are there in 6 and one fourth?

6. How many whole things are there in 2 and 5 fourths, and 5 and 9 fourths?

7. If a barrel of vinegar cost 4 dollars, what will 5 and 3 fourths barrels cost?

8. If 9 fourths yards of muslin cost 18 cents, what will 12 yards cost?

9. If 11 fourths yards of broadcloth cost 22 dollars what will 7 yards cost?

10. If 3 and a fourth yards of cambric cost 39 cents, how much will 7 yards cost?

LESSON IV.

Of the Fraction One Fifth.

1. If an apple be divided into 5 equal parts, one of the parts is called one fifth. What are 2 of them called? 3 of them? 4 of them? 5 of them?

2. How many fifths are there in 1 apple? How many fifths in 1 thing?

3. How many fifths are there in 2? In 2 and one fifth? In 2 and 2 fifths?

4. How many fifths are there in 3? In 3 and one fifth? In 3 and two fifths? In 3 and three fifths?

5. How many fifths are there in 4? In 4 and one fifth? In 4 and two fifths? In 4 and three fifths?

6. How many fifths are there in 5? In 5 and three fifths? In 5 and one fifth? In 5 and four fifths?

7. How many fifths are there in 6? In 6 and one fifth? In 6 and two fifths? In 6 and four fifths?

8. How many fifths are there in 7? In 7 and one fifth? In 7 and two fifths? In 7 and three fifths? In 7 and four fifths?

9. How many fifths are there in 8? In 8 and two fifths? In 8 and three fifths? In 8 and one fifth?

10. How many fifths are there in 9? In 9 and one fifth? In 9 and three fifths? In 9 and four fifths?

11. How many fifths are there in 10? In 10 and one fifth? In 10 and three fifths? In 10 and four fifths?

12. How many whole apples are equal to 5 fifths of an apple? How many whole things in 5 fifths?

13. How many whole things in 10 fifths? In 12 fifths? In 14 fifths? In 11 fifths?

14. How many whole things in 15 fifths? In 18 fifths? In 25 fifths? In 37 fifths? In 40 fifths? In 60 fifths?

15. How many fifths in 8 and 4 fifths? In 6 and 3 fifths? In 4 and 2 fifths? In 11 and 4 fifths?

QUESTIONS.

1. If 1 fifth of a yard of broadcloth cost 1 dollar, what will 1 yard cost?

2. If 3 fifths of a pound of coffee cost 15 cents, what will 3 pounds cost?

3. If 4 fifths of a pound of tea cost 40 cents, what will 4 pounds cost?

4. If 1 yard of cloth cost 3 fifths of a dollar, what will 6 yards cost?

5. If 1 pound of tea cost 3 fifths of a dollar, what will 5 pounds cost?

6. If 2 yards of cloth cost 4 fifths of a dollar, what will 10 yards cost?

7. John has 9 fifths apples, James 6 fifths, Charles 7 fifths, and Reuben 3 fifths: how many entire apples have they in all?

8. If my pencil cost 4 fifths of a dime, my pen 1 fifth, my paper 3 fifths, and wafers 2 fifths: how much do I pay in all?

9. If a family consume 3 fifths of a barrel of flour in 1 week, how much will they consume in 5 weeks?

10. Eight times 3 fifths are how many fifths? How many times 1?

LESSON V.

Of the Fraction One Sixth.

1. If an apple be divided into six equal parts, one of the parts is called one sixth: what are 2 of them called? 3 of them? 4 of them? 5 of them? 6 of them?

2. How many sixth parts are there in 1 apple?

3. How many sixths are there in 1?

4. How many sixths are there in 1 and one sixth?

5. How many sixths are there in 2? In 2 and 3 sixths? In 2 and 4 sixths?

6. How many sixths are there in 3? In 3 and 4 sixths? In 3 and 5 sixths? In 3 and 2 sixths?

7. How many sixths are there in 4? In 4 and 3 sixths? In 4 and 5 sixths? In 4 and 2 sixths?

8. How many sixths are there in 5? In 5 and 3 sixths? In 5 and 4 sixths? In 5 and 5 sixths?

9. How many sixths are there in 6? In 6 and 3 sixths? In 6 and 5 sixths?

10. How many sixths are there in 7? In 7 and 1 sixth? In 7 and 2 sixths? In 7 and 5 sixths?

11. How many sixths are there in 8? In 8 and 1 sixth? In 8 and 2 sixths?

12. How many sixths are there in 9? In 9 and 2 sixths? In 9 and 4 sixths?

13. How many sixths are there in 10? In 10 and 1 sixth? In 10 and 2 sixths? In 10 and 3 sixths? In 10 and 4 sixths?

14. How many whole apples are there in 6 sixths apples? In 12 sixths? In 18 sixths? In 24 sixths? In 36 sixths? In 72 sixths?

15. How many whole things in 15 sixths? In 54 sixths? In 66 sixths? In 45? In 50? In 48? In 63? In 75?

QUESTIONS.

1. If a family consume 1 bushel and 3 sixths of a bushel of potatoes in 1 week, how much will they consume in 8 weeks?

2. How many sixths of apples can you cut from apples and 2 sixths? How many from 6 and 5 sixths?

3. If James gives 5 sixths of a shilling for a pencil, and 4 sixths of a shilling for a slate, and 3 sixths of a shilling for paper: how much will he pay in all?

4. If 2 sixths of a barrel of flour will last a family 4 weeks, how long will 2 barrels last them?

5. If 4 sixths of a load of hay is worth 8 dollars, what are 4 loads worth, at the same rate?

6. What is the cost of 3 barrels of flour at 4 dollars and 2 sixths a barrel?

7. A man gave to each of 12 men 5 sixths of a loaf of bread: how many loaves did he distribute?

8. A man bought 6 barrels of flour at 5 and 5 sixths dollars a barrel: what did it come to?

9. What is the product of 5 and 4 sixths, multiplied by 9?

10. What is the product of 6 and 5 sixths, multiplied by 12?

LESSON VI.

Of the Fraction One Seventh and One Eighth.

1. If an apple be divided into 7 equal parts, one of the parts is called one seventh: what are 2 of the parts called? 3 of them? 4 of them? 5 of them? 6 of them? 7 of them?

How many sevenths are there in 1 apple? How many sevenths in 1 thing?

2. How many sevenths are there in 2? In 2 and 1 seventh, how many? In 2 and 2 sevenths, how many?

3. How many sevenths are there in 3? In 3 and 1 seventh, how many? In 3 and 2 sevenths, how many?

4. How many sevenths are there in 4? In 4 and 3 sevenths, how many? In 4 and 4 sevenths, how many? In 4 and 6 sevenths?

5. How many sevenths are there in 5? In 5 and 2 sevenths, how many? In 5 and 3 sevenths, how many? In 5 and 6 sevenths?

6. How many sevenths are there in 6? In 6 and 2 sevenths, how many? In 6 and 3 sevenths? In 6 and 4 sevenths?

7. How many sevenths are there in 7? In 7 and 1 seventh? In 7 and 3 sevenths? In 7 and 4 sevenths?

8. How many sevenths are there in 8? In 8 and 1 seventh? In 8 and two sevenths? In 8 and 5 sevenths?

9. How many sevenths are there in 9? In 9 and 4 sevenths? In 9 and 3 sevenths?

10. How many sevenths are in 10? In 10 and 1 seventh? In 10 and 2 sevenths? In 10 and 3 sevenths?

11. How many whole things in 14 sevenths? In 28 sevenths? In 35 sevenths? In 49 sevenths?

12. How many whole things in 17 sevenths? In 14 sevenths and 4 sevenths? How many in 63 sevenths?

13. How many sevenths in 9 and 5 sevenths?

14. How many whole things in 30 sevenths, 14 sevenths and 5 sevenths?

EIGHTHS.

1. If an apple be divided into eight equal parts, what is one of the parts called? How many eighths are there in one? What are two parts called? 3 parts? 4 parts? 5 parts? 7 parts?

2. How many eighths are there in 2? How many in 2 and 1 eighth? In 2 and 5 eighths? In 2 and 6 eighths?

3. How many eighths are there in 3? In 3 and 2 eighths? In 3 and 4 eighths? In 3 and 7 eighths? In 3 and 6 eighths?

4. How many eighths in 4? In 4 and 1 eighth? In 4 and 3 eighths? In 4 and 5 eighths? In 4 and 6 eighths?

5. How many eighths are there in 5? In 5 and 1 eighth? In 5 and 2 eighths? In 5 and 3 eighths? In 5 and 5 eighths?

6. How many eighths are there in 6? In 6 and 2 eighths? In 6 and 3 eighths? In 6 and 7 eighths? In 6 and 4 eighths?

7. How many eighths are there in 7? In 7 and 1 eighth? In 7 and 2 eighths? In 7 and 3 eighths?

8. How many eighths are there in 8? In 8 and 1 eighth? In eight and 3 eighths? In 8 and 4 eighths? In 8 and 6 eighths?

9. How many eighths are there in 9? In 9 and 1 eighth? In 9 and 2 eighths? In 9 and 3 eighths? In 9 and 4 eighths?

10. How many eighths are there in 10? In 10 and 1 eighth? In 10 and 2 eighths? In 10 and 3 eighths?

11. How many whole things are there in 16 eighths? In 24 eighths, how many? How many in 32 eighths? In 40? In 50? In 60? In 70? In 56? In 96? In 88? In 26?

LESSON VII.

Of the Fractions One Ninth and One Tenth.

1. If an apple be divided into 9 equal parts, one of the parts is called one ninth. How many ninths are there in 1? What are 2 of the parts called? 5 of them? 7 of them?

2. How many ninths are there in 2 things? How many in 2 and 2 ninths? In 2 and 4 ninths? In 2 and 5 ninths? In 2 and 6 ninths?

3. How many ninths are there in 3? In 3 and 7 ninths? In 3 and 6 ninths? In 3 and 5 ninths? In 3 and 8 ninths?

4. How many ninths are there in 4 things? In 4 and 1 ninth? In 4 and 3 ninths? In 4 and 6 ninths? In 4 and 8 ninths?

5. How many ninths are there in 5? In 5 and 2 ninths? In 5 and 3 ninths? In 5 and 4 ninths? In 5 and 6 ninths?

6. How many ninths are there in 6? In 6 and 4 ninths? In 6 and 5 ninths? In 6 and 8 ninths?

7. How many ninths are there in 7? In 7 and 3 ninths? In 7 and 4 ninths? In 7 and 6 ninths? In 7 and 8 ninths?

8. How many ninths are there in eight? In 8 and 1 ninth? In 8 and 2 ninths? In 8 and 4 ninths?

9. How many ninths are there in 9? In 9 and 3 ninths? In 9 and 4 ninths? In 9 and 5 ninths?

10. How many ninths are there in 10? In 10 and 1 ninth? In 10 and 2 ninths? In 10 and 8 ninths?

11. How many whole things in 18 ninths? In 27? In 36? In 63? In 72? In 90? In 81? In 54? In 67? In 59? In 71? In 99?

TENTHS.

1. If an apple be divided into ten equal parts, one of the parts is called one tenth. How many tenths are there in one thing? What are 2 parts called? 4? 5? 7?

2. How many tenths are there in 2? In 2 and 3 tenths? In 2 and 5 tenths? In 2 and 9 tenths?

3. How many tenths are there in 3? In 3 and 4 tenths? In 3 and 5 tenths? In 3 and 6 tenths?

4. How many tenths are there in 4? In 4 and 4 tenths? In 4 and 5 tenths? In 4 and 8 tenths?

5. How many tenths are there in 5? In 5 and 3 tenths? In 5 and 6 tenths? In 5 and 9 tenths?

6. How many tenths are there in 6? In 6 and 3 tenths? In 6 and 7 tenths? In 6 and 8 tenths?

7. How many tenths are there in 7? In 7 and 3 tenths? In 7 and 8 tenths? In 7 and 9 tenths?

8. How many tenths are there in 8? In 8 and 4 tenths? In 8 and 5 tenths? In 8 and 9 tenths?

9. How many tenths are there in 9? In 9 and 4 tenths? In 9 and 5 tenths? In 9 and 8 tenths?

10. How many tenths are there in 10? In 10 and 5 tenths? In 10 and 6 tenths? In 10 and 9 tenths?

11. How many whole things in 20 tenths? In 55? In 60? In 70? In 80? In 85? In 95? In 100?

LESSON VIII.

Equal parts of Numbers.

1. What is one half of 4?
2. What part of 1 is one third of 2?*
3. What part of 1 is one fourth of 3?
4. What part of 1 is one ninth of 6?
5. What part of 1 is one fifth of 4?
5. How many times 1 is one sixth of 8?
7. How many times 1 is one fifth of 12?
8. Nine are how many times 7?
9. If wheat is 9 shillings a bushel, how many bushels can you buy for 10 shillings? For 15 shillings? For 20 shillings? For 29 shillings?
10. Twelve are how many times 6?
11. What part of 1 is one twelfth of 6?
12. What part of 1 is one tenth of 4?
13. What part of 1 is one twelfth of 9?
4. At 9 dimes a yard, how much muslin can you buy for 1 dime? How much for 4 dimes? How much for 7? How much for 9?

* SUGGESTION.—One third of 2 must be twice as great as one third of 1: one third of one is one third; hence, one third of 2 is two thirds of 1. Similarly for each example.

OBSERVE that all numbers are finally expressed in terms of the unit 1.

How many for 12? How many for 18? How many for 20? How many for 27?

15. If coal is 7 dollars a ton, how much will one seventh of a ton cost? 5 sevenths? 14 sevenths?

16. If you have 30 dollars, how many barrels of flour can you buy at 5 dollars a barrel? At 6 dollars a barrel? At 7 dollars? At 8 dollars? At 9 dollars? At 10 dollars?

17. Twenty-four are how many times 4? How many times 5? 8? 9? 10? 11? 12?

18. Sixty are how many times 5? 7? 8? 4? 9? 10? 12?

19. Seventy are how many times 10? 7? 8? 9? 11? 3?

20. Forty-two are how many times 8? 9? 10? 12? 6? 7?

21. Fifty-six are how many times 12? 10? 9? 8? 11? 5?

22. Twenty-two are how many times 5? 6? 7? 8? 9? 10?

23. Forty-eight are how many times 7? 6? 5? 4? 12? 11? 9?

24. Fifty-five are how many times 8? 9? 7? 5? 3? 10? 11?

25. Forty-seven are how many times 5? 6? 9? 10? 12? 8? 7?

26. Forty-one are how many times 6? 7? 9? 8? 5? 4?

27. If cloth is 7 dollars a yard, how many yards can you buy for 60 dollars? How much could you buy at 8 dollars a yard? At 10 dollars? 5 dollars? 11 dollars?

28. Eighty-two are how many times 4? 8? 9? 12? 11?

29. Eighty-eight are how many times 7? 8? 9? 10? 11? 12?

30. Eighty-five are how many times 9? 8? 7? 10? 11? 12?

31. Eighty-seven are how many times 9? 8? 7? 10? 11? 12?

32. Ninety are how many times 9? 8? 7? 10? 11? 12?

33. Ninety-nine are how many times 9? 11? 12? 8? 10?

34. Seventy-two are how many times 6? 9? 7? 8? 10? 11? 12?

35. Seventy-six are how many times 9? 8? 7? 6? 10? 11? 12?

36. Seventy-eight are how many times 9? 8? 7? 10? 12? 11?

37. At 7 dollars a barrel, how much flour can you buy for 2 dollars? How much for 4 dollars? How much for 6 dollars? For 8 dollars? For 12 dollars? For 15 dollars? For 20 dollars? For 25 dollars? For 27 dollars? For 32 dollars? For 35 dollars?

38. Fifty-five are how many times 5? 6? 8? 9? 10? 12? 11?

39. Fifty-nine are how many times 8? 7? 6? 9? 10? 11? 12?

40. If sugar is 9 dollars a hundred weight, how many hundreds can you buy for 3 dollars? How much for 6? For 12? For 18 dollars? For 26? For 39? For 65? For 72? For 92 dollars?

LESSON IX.

Expressing Fractions by Figures and Reading them.

1. The following is the manner of expressing fractions by figures:

$\frac{1}{2}$	one half.	$\frac{1}{7}$	one seventh.
$\frac{1}{3}$	one third.	$\frac{1}{8}$	one eighth.
$\frac{1}{4}$	one fourth.	$\frac{1}{9}$	one ninth.
$\frac{1}{5}$	one fifth.	$\frac{1}{10}$	one tenth.
$\frac{1}{6}$	one sixth.	$\frac{1}{12}$	one twelfth.

2. What is the figure above the line called? The numerator.

3. What is the figure below the line called? The denominator.

4. What does the denominator show?

It shows into how many equal parts the whole thing is divided.

5. What does the numerator show?

How many of those equal parts are taken.

6. What may the whole thing which is divided be called? Unity or one.

7. Is one apple a unit? Is one peach a unit? Is one dollar a unit? Is one book a unit?

8. In the fraction $\frac{1}{3}$, into how many equal parts is the unit divided? Which figure is the numerator? Which the denominator?

9. In the fraction $\frac{1}{12}$, into how many equal parts is the unit divided? Which is the numerator? Which the denominator?

10. Read the following fractions:—

$\frac{3}{8}$	three eighths.	$\frac{8}{13}$	eight thirteenths.
$\frac{3}{7}$	three sevenths.	$\frac{41}{16}$	forty-one sixteenths.
$\frac{5}{9}$	five ninths.	$2\frac{3}{8}$	two and 3 eighths.
$\frac{6}{12}$	six twelfths.	$5\frac{9}{7}$	five and 9 sevenths.
$\frac{7}{15}$	seven fifteenths.	$6\frac{5}{12}$	six and 5 twelfths.*

11. In the fraction $\frac{3}{8}$, into how many equal parts is the unit divided? How many of these parts are taken?

12. In the fraction $\frac{3}{7}$, into how many equal parts is the unit divided? How many parts are taken?

---

* SUGGESTIONS.—When the numerator is less than the denominator, the fraction is called a *proper fraction.*

When the numerator is greater than the denominator, the fraction is called an *improper fraction.*

A whole number united with a fraction, is called a *mixed number.*

13. In the fraction $\frac{5}{8}$, into how many equal parts is the unit divided? How many parts are taken?

14. In the fraction $\frac{5}{9}$, into how many equal parts is the unit divided? How many parts are taken?

15. In the fraction $\frac{6}{12}$, into how many equal parts is the unit divided? How many parts are taken?

16. In the fraction $\frac{7}{15}$, into how many equal parts is the unit divided? How many parts are taken?

17. In the fraction $\frac{8}{13}$, into how many equal parts is the unit divided? How many parts are taken?

18. In the fraction $\frac{27}{18}$, into how many equal parts is the unit divided? How many parts are taken?

19. What figures express one third of one?

20. What figures express five sevenths?

21. What figures express nine twelfths?

22. What figures express twelve seventeenths?

23. What figures express ten thirty-sevenths?

24. What figures express fifteen sevenths?

25. What figures express fourteen twenty-ninths?

LESSON X.

Fractional Units.

1. What is the unit of a fraction?

It is the whole thing which is divided into equal parts.

2. What is each equal part called?

A *fractional unit.*

3. In the fraction $\frac{5}{8}$ of a pound, what is the unit of the fraction?

One pound.

What is the fractional unit?

One-eighth of a pound.

How many fractional units are taken?

Five.

4. In the fraction $\frac{3}{4}$, what is the unit of the fraction?

The abstract or simple unit, one.
What is the fractional unit?
One fourth.
How many fractional units are taken?
Three.

5. What is the fractional unit in three fourths?
How many fractional units in three fourths?

6. What is the unit of the fraction $\frac{3}{7}$ of a dollar?
What is the fractional unit?
How many fractional units are taken?

7. What is the unit of the fraction $\frac{3}{5}$?
What is the fractional unit?
How many are taken?

8. How many fractional units in $\frac{11}{12}$?
What is the unit of the fraction?
What the fractional unit?

9. James and John have each an apple of the same size. James cuts his into two equal parts, and gives away one half. John cuts his into three equal parts and gives away two of the pieces. They seek to find what part of a whole apple they have left.*

---

* SUGGESTION.—This example suggests all the principles employed in fractions. They may be thus stated:

1st. That something regarded as a whole, called unity, is the primary base of every fraction: and

2d. That one of the equal parts of unity, called the fractional unit, is the second base of any fractional number.

From the nature of Division and Multiplication, we see:

1st. That the fractional unit is as many times less than unity as there are units in the denominator:

2d. That the numerator shows how many fractional units are taken.

3d. If the numerator be multiplied by any number, the number of fractional units will be increased as many times as there are units in the multiplier.

4th. If the numerator be divided by any number, the number of fractional units will be diminished as many times as there are units in the divisor.

Now, says James, if I cut my half into 3 equal parts, each part will be three times less than before; that is, each part will be one sixth of the entire apple, but I shall have three times as many parts, so that I shall still have half my apple.

Now, says John, if I cut my third into 2 equal arts, each part will be two times less than before; hat is, it will be one-sixth of the entire apple; but I shall have twice as many parts, so that I still have cne third of the apple.

Now, says John to James, if you have 3 sixths and I have 2 sixths, together we must have 5 sixths of a whole apple.

10. In the fractions $\frac{1}{2}$ and $\frac{1}{3}$, what is the fractional unit of the first? What of the second?

How can you reduce them to the same fractional unit?

By multiplying the numerator and denominator of the first by 3, and of the second by 2.

What is the fractional unit after the multiplication?

How many fractional units are taken in each fraction? How many in both?

11. How can you reduce fractions to the same fractional unit?

By multiplying the numerator and denominator of

5th. If the denominator be multiplied by any number, the fractional unit will be diminished as many times as there are units in the multiplier.

6th. If the denominator be divided by any number, the value of the fractional unit will be increased as many times as there are units in the divisor.

By combining 3 and 5, we see:

7th. That if the numerator and denominator be multiplied by the same number, the value of the fraction will not be changed.

And by combining 4 and 6, we see:

8th. That if the numerator and denominator be both divided by the same number, the value of the fraction will not be changed.

each fraction by such a number as shall make the denominators the same in all.

12. Reduce $\frac{1}{3}$ and $\frac{1}{4}$, to the same fractional unit?*

If we multiply the numerator and denominator of the first fraction by 4, we have $\frac{4}{12}$; and if we multiply the second by 3. we have $\frac{3}{12}$, in which the fractional unit is $\frac{1}{12}$ in both.

13. Reduce $\frac{1}{4}$ and $\frac{1}{5}$ to the same fractional unit.

After reduction, what is the fractional unit, and how many fractional units in both?

14. Reduce $\frac{1}{5}$ and $\frac{1}{6}$ to the same fractional unit.

After reduction, what is the fractional unit, and how many fractional units in both?

15. Reduce $\frac{1}{2}$, $\frac{2}{3}$, and $\frac{3}{4}$, to the same fractional unit.

After reduction, how many fractional units in each?

16. Reduce $\frac{3}{4}$ and $\frac{4}{5}$ to the same fractional unit.

After reduction, what is the fractional unit?

How many in each?

17. Ruduce $\frac{5}{8}$. $\frac{1}{2}$, and $\frac{2}{3}$, to the same fractional unit.

18. Reduce $\frac{3}{5}$, $\frac{4}{3}$, and $\frac{2}{6}$. to the same fractional unit.

19. Reduce $\frac{4}{3}$, $\frac{3}{6}$, and $\frac{2}{8}$, to the same fractional unit.

20. Reduce $\frac{3}{5}$ and $\frac{4}{9}$ to the same fractional unit.

21. Reduce $\frac{1}{6}$ and $\frac{5}{8}$ to the same fractional unit.

22. Reduce $\frac{4}{9}$ and $\frac{5}{8}$ to the same fractional unit.

23. Reduce $\frac{1}{2}$, $\frac{4}{5}$, and $\frac{3}{6}$, to the same fractional unit.

24. In $\frac{15}{3}$ how many units 1?

25. In $\frac{67}{9}$ how many units 1?

26. In $\frac{45}{9}$ how many units 1?

27. In $\frac{36}{6}$ how many units 1?

28. In $\frac{47}{5}$ how many units 1?

29. In $\frac{24}{8}$ how many units 1?

---

* SUGGESTIONS.—There are but three operations which chang the units of numbers: They are,

1st. To change integral to fractional units:

2d. To change fractional to integral units: and

3d. To change from one fractional unit to another.

The first two were fully explained in the first seven lessons of this section: and the third is treated of in this.

30. In $\frac{27}{5}$ how many units 1?
31. In $\frac{39}{8}$ how many units 1?
32. In $\frac{54}{9}$ how many units 1?
33. In $\frac{75}{9}$ how many units 1?
34. How many units 1 in $\frac{37}{5}$?
35. How many units 1 in $\frac{49}{7}$?
36. How many nnits 1 in $\frac{63}{7}$?
37. How many units 1 in $\frac{49}{7}$?
38. How many fourths in 1 half?
39. How many fourths in 2 halves?
40. How many fourths in 6 halves?
41. How many thirds in 12 sixths?
42. How many sixths in 1 third?
43. How many sixths in 4 thirds?
44. How many sixths in 1 half?
45. How many sixths in 2 halves?
46. How many eighths in 1 fourth?
47. How many eighths in 1 half?
48. How many eighths in 2 fourths?
49. How many eighths in 3 fourths?
50. How many ninths in 1 third?
51. How many ninths in 2 thirds?
52. How many ninths in 6 thirds?
53. How many tenths in 1 fifth?
54. How many tenths in 1 half?
55. How many twelfths in 1 third?
56. How many twelfths in 1 half?
57. How many twelfths in 1 quarter?
58. How many sixteenths in 1 eighth?
59. How many sixteenths in 1 quarter?
60. How many eighteenths in 1 ninth?
61. How many eighteenths in 1 half?
62. How many eighteenths in 2 thirds?
63. How many twentieths in 3 halves?
64. How many twentieths in 3 quarters?
65. How many twentieths in 5 fourths?
66. How many twentieths in 3 fifths?

67. How many halves in $\frac{8}{16}$? In $\frac{6}{12}$?*
68. How many halves in $\frac{24}{12}$? In $\frac{6}{4}$?
69. How many thirds in $\frac{3}{9}$? In $\frac{4}{12}$?
70. How many thirds in $\frac{8}{12}$? In $\frac{4}{12}$?
71. How many thirds in $\frac{9}{27}$? In $\frac{32}{48}$?
72. How many fourths in $\frac{12}{16}$? In $\frac{16}{32}$?
73. How many fourths in $\frac{30}{24}$? In $\frac{4}{16}$?
74. How many fourths in $\frac{72}{36}$? In $\frac{18}{36}$?
75. How many fourths in $\frac{10}{40}$? In $\frac{48}{16}$?
76. How many fifths in $\frac{6}{10}$? In $\frac{8}{10}$?
77. How many fifths in $\frac{20}{50}$? In $\frac{6}{30}$?
78. How many sixths in $\frac{36}{54}$? In $\frac{8}{48}$?
79. How many sixths in $\frac{72}{36}$? In $\frac{32}{48}$?
80. How many sevenths in $\frac{14}{49}$? In $\frac{48}{56}$?
81. How many sevenths in $\frac{21}{63}$? In $\frac{6}{14}$?

LESSON XI.

Value of Numbers.

1. What is the base of a number?

The primary base of every number is the UNIT ONE.

2. What is an integer, or whole number?

It is a number which contains the unit one an exact number of times. Thus, three, four, five, &c., are integers, or whole numbers.

3. What is a fractional number?

It is a number which expresses one or more of the equal parts of unity. Thus, $\frac{1}{2}$, $\frac{3}{4}$, $\frac{5}{8}$, &c., are fractions.

4. What expresses the value of a number either integral or fractional?

The number of times which it contains the unit one.

5. How many times does five contain one?

6. How many times is five greater than one? Why?

---

* SUGGESTIONS.—Divide the numerator and denominator by such a number as will give the required fractional unit.

7. How many times is six greater than one? Why?

8. How many times is eight greater than one? Why?

9. How many times does one half contain the unit one?

One-half times.

10. How many times is one half less than one?

Two times.

11. What expresses the value of a fraction?

The number of times which a fraction contains 1.

12. What is the value of one half? Why?

13. What is the value of two halves? Why?

14. What is the value of three halves? Why?

15. What is the value of four halves? Why?

16. What is the value of one third? Why?

17. What is the value of three thirds? Why?

18. What is the value of six thirds? Why?

19. How do you find the value of an improper fraction?*

By dividing the numerator by the denominator.

20. If there is a remainder, what do you do with it?

Write the denominator under it, and annex the fraction to the integral number.

21. What is the value of the fraction seven thirds?

22. What is the value of nine fourths? Of seventy-two twelfths? Of one hundred eighths? Of ninety-four ninths?

23. What is the value of seventy-five twelfths? Of sixty-seven eighths? Of eighty-nine elevenths? Of one hundred and twenty twelfths?

---

* ANALYSIS.—Since the numerator shows how many fractional units are taken, and the denominator how many fractional units make 1, it follows, that the numerator divided by the denominator will show how many units 1, there are in the fraction.

Impress the pupil, constantly, that every number, whether integral or fractional, must be compared with the unit one. Also, that the value of any number is expressed by the number of times which it contains the unit one.

24. When is a fraction said to be in its lowest terms?

A fraction is said to be in its lowest terms when there is no number except 1 that will divide both the numerator and denominator without a remainder.

25. How may a fraction be reduced to its lowest terms?

By dividing both the numerator and denominator by the same number.*

26. Is the fraction $\frac{3}{5}$ in its lowest terms? Why?

27. What fraction expresses the lowest terms of $\frac{4}{8}$?

28. What fraction expresses the lowest terms of $\frac{8}{10}$?

29. What fraction expresses the lowest terms of $\frac{6}{16}$?

30. What fraction expresses the lowest terms of $\frac{4}{12}$?

31. What are the lowest terms of the fraction $\frac{6}{12}$?

32. What are the lowest terms of the fraction $\frac{3}{7}$?

33. What are the lowest terms of the fraction $\frac{5}{15}$?

34. What are the lowest terms of the fraction $\frac{8}{24}$?

35. What are the lowest terms of the fraction $\frac{6}{24}$?

36. What are the lowest terms of the fraction $\frac{16}{32}$?

37. Reduce the fraction $\frac{16}{48}$ to its lowest terms.

38. What fraction will express the lowest terms of $\frac{24}{48}$?

39. What fraction will express the lowest terms of $\frac{18}{30}$?

40. What fraction will express the lowest terms of $\frac{22}{36}$?

SUGGESTION.—*Reduce every fraction to its lowest terms before performing any other operation. Reduce also every answer to its lowest terms.*

LESSON XII.

Adding Fractional Units.

James and John have each an apple of the same size. James cuts his into 4 equal parts and gives away 2 parts. John cuts his into 5 equal parts and gives away 3 parts. They seek to find what part of an apple each has left, and what part both together have?

Now, says James, if I cut each of my fourths into 5 equal parts, I shall have 10 parts; that is, 10 twentieths of an apple: and says John, if I cut each of my fifths into 4 equal parts, I shall have 8 parts; that is, 8 twentieths of an apple; hence, James has 10 twentieths, and John 8 twentieths, and together they have 18 twentieths.

2. What is the unit of the fraction $\frac{18}{20}$? What is the fractional unit in 18 twentieths of an apple? How many fractional units are taken? If the numerator and denominator be each divided by 2, what does the fraction become? What, then, is each fractional unit? How many are taken?

3. What is necessary in order that two or more fractions may be added together.

That they have the same integral unit, and the same fractional unit.

4. When no integral unit is named, what unit is understood?

The abstract unit one.

5. Have the fractions $\frac{1}{4}$ and $\frac{2}{4}$ the same integral unit? What is it? Have they the same fractional unit? What is it?

6. How many fractional units in the first? How many in the second? How many in both?

7. What is the sum of the fractional units in $\frac{1}{2}$ and $\frac{2}{2}$?

8. What is the sum of $\frac{3}{4}$ and $\frac{2}{4}$?

9. What is the sum of $\frac{1}{2}$, $\frac{2}{2}$, and $\frac{3}{2}$?
10. What is the sum of $\frac{1}{3}$, $\frac{4}{3}$, and $\frac{5}{3}$?
11. What is the sum of $\frac{4}{7}$, $\frac{5}{7}$, and $\frac{6}{7}$?
12. What is the sum of $\frac{5}{8}$, $\frac{3}{8}$, $\frac{4}{8}$, and $\frac{2}{8}$?
13. What is the sum of $\frac{5}{9}$, $\frac{3}{9}$, and $\frac{4}{9}$?
14. What is the sum of $\frac{3}{7}$, $\frac{5}{7}$, and $\frac{6}{7}$?
15. What is the sum of $\frac{3}{4}$, $\frac{5}{4}$, $\frac{6}{4}$, and $\frac{7}{4}$?
16. What is the sum of $\frac{1}{2}$, $\frac{3}{2}$, $\frac{6}{2}$, and $\frac{5}{2}$?
17. What is the sum of $\frac{3}{3}$, $\frac{4}{3}$, $\frac{5}{6}$, and $\frac{6}{3}$?
18. What is the sum of $\frac{6}{2}$, $\frac{7}{2}$, $\frac{8}{2}$, and $\frac{9}{2}$?
19. What is the sum of $\frac{3}{4}$, $\frac{4}{4}$, $\frac{6}{4}$, and $\frac{7}{4}$?
20. What is the sum of $\frac{4}{5}$, $\frac{3}{5}$, $\frac{2}{5}$, and $\frac{1}{5}$?
21. What is the sum of $\frac{2}{6}$, $\frac{3}{6}$, $\frac{5}{6}$, and $\frac{7}{6}$?
22. What is the sum of $\frac{13}{7}$, $\frac{4}{7}$, $\frac{3}{7}$, and $\frac{1}{7}$?
23. What is the sum of $\frac{3}{8}$, $\frac{9}{8}$, $\frac{7}{8}$, and $\frac{2}{8}$?
24. What is the sum of $\frac{1}{9}$, $\frac{2}{9}$, $\frac{10}{9}$, and $\frac{12}{9}$?
25. What is the sum of $\frac{1}{2}$, $\frac{3}{2}$, $\frac{5}{2}$, $\frac{6}{2}$, and $\frac{7}{2}$?
26. What is the sum of $\frac{5}{3}$, $\frac{6}{3}$, $\frac{9}{3}$, and $\frac{8}{3}$?
27. What is the sum of $\frac{8}{4}$, $\frac{3}{4}$, $\frac{5}{4}$, and $\frac{7}{4}$?
28. What is the sum of $\frac{3}{5}$, $\frac{4}{5}$, $\frac{7}{5}$, and $\frac{9}{5}$?
29. What is the sum of $\frac{5}{6}$, $\frac{6}{6}$, $\frac{9}{6}$, and $\frac{10}{6}$?
30. Add together $\frac{1}{2}$ and $\frac{1}{3}$.

First, $\frac{1}{2}$ is equal to $\frac{3}{6}$, and $\frac{1}{3}$ is equal to $\frac{2}{6}$; then $\frac{3}{6}$ plus $\frac{2}{6}$ are equal to $\frac{5}{6}$.

31. What is the sum of $\frac{3}{2}$ and $\frac{3}{4}$?
32. What is the sum of $\frac{3}{5}$ and $\frac{4}{6}$? (Reduce to thirtieths.)
33. What is the sum of $\frac{2}{7}$ and $\frac{3}{14}$?
34. What is the sum of $\frac{3}{8}$ and $\frac{6}{24}$?
35. What is the sum of $\frac{4}{9}$ and $\frac{3}{18}$?
36. What is the sum of $\frac{2}{3}$ and $\frac{5}{12}$?
37. What is the sum of $\frac{5}{8}$ and $\frac{3}{2}$?
38. What is the sum of $\frac{4}{5}$ and $\frac{3}{2}$?
39. What is the sum of $\frac{5}{7}$ and $\frac{4}{3}$?
90. What is the sum of $\frac{4}{9}$ and $\frac{5}{6}$?
41. What is the sum of $\frac{5}{8}$ and $\frac{3}{4}$?

42. What is the sum of $2\frac{3}{4}$ and $3\frac{1}{4}$ and $5\frac{1}{12}$?*

43. What is the sum of $2\frac{1}{3}$ and $4\frac{4}{3}$ and $6\frac{1}{6}$?

44. What is the sum of $1\frac{2}{5}$ and $6\frac{4}{5}$ and $7\frac{6}{20}$?

45. What is the sum of $4\frac{2}{9}$ and $\frac{5}{9}$ and $2\frac{3}{27}$?

46. What is the sum of $9\frac{4}{8}$ and $\frac{5}{8}$ and $7\frac{6}{16}$?

47. What is the sum of $5\frac{3}{7}$ and $4\frac{5}{7}$ and $\frac{5}{28}$?

48. What is the sum of $9\frac{4}{3}$ and $7\frac{5}{3}$ and $\frac{7}{18}$?

49. What is the sum of $5\frac{8}{5}$ and $4\frac{9}{15}$ and $8\frac{9}{30}$?

50. What is the sum of $6\frac{4}{9}$ and $8\frac{6}{18}$ and $10\frac{9}{27}$?

51. What is the sum of $5\frac{3}{7}$ and $8\frac{8}{14}$ and $2\frac{9}{21}$?

52. What is the sum of 5 and $2\frac{1}{2}$ and $3\frac{1}{6}$?

QUESTIONS.

1 John buys a top for one sixth of a shilling, a stick of candy for one twelfth of a shilling, and a piece of india-rubber for one third of a shilling: what does the whole cost him?†

2. James pays $\frac{3}{7}$ of a dollar for a pair of gloves, and $\frac{3}{8}$ of a dollar for a handkerchief: how much do they cost him?

3. Nancy buys a work-box for $\frac{7}{8}$ of a dollar, a pair of gloves for $\frac{3}{8}$ of a dollar, and a comb for $\frac{2}{16}$ of a dollar: how much do they all cost?

4. Jane buys a yard of ribbon for $\frac{2}{7}$ of a dollar, a gold pin for $\frac{5}{2}$ of a dollar, and an inkstand for $\frac{3}{14}$ of a dollar: how much did she pay in all?

5. William buys a kite for $\frac{2}{3}$ of a dollar, and a string for $\frac{4}{15}$ of a dollar: how much did he pay?

6. Three ducks cost $\frac{4}{5}$ of a dollar, two fowls $\frac{2}{3}$ of a dollar, and two geese $\frac{9}{15}$ of a dollar: what is the entire cost?

---

* SUGGESTIONS.—Add the integer numbers separately, and unite the sum to the sum of the fractional parts.

† ANALYSIS.— $\frac{1}{6}$ of a shilling is $\frac{2}{12}$ of a shilling.
$\frac{1}{12}$ of a shilling is $\frac{1}{12}$ "
$\frac{1}{3}$ of a shilling is $\frac{4}{12}$ "
hence the whole costs is - - - $\frac{7}{12}$ of a shilling.

7. Three sheep cost $\frac{17}{2}$ of a dollar, a calf $\frac{9}{4}$ of a dollar, and a lamb $\frac{5}{4}$ of a dollar: what is the entire cost?

8. Three yards of cloth cost $\frac{4}{9}$ of a dollar, a handkerchief $\frac{7}{9}$ of a dollar, and a pair of gloves $\frac{1}{3}$ of a dollar: what is the entire cost?

9. A father paid 3 quarters of a dollar for his own breakfast, one third of a dollar for his son's, and a quarter of a dollar for his daughter's: how much did he pay in all?

10. A merchant sold $3\frac{1}{2}$ yards of cloth from one piece, $2\frac{1}{3}$ yards from another, and $5\frac{1}{4}$ yards from another: how much did he sell in all?*

11. If a turkey costs $\frac{7}{8}$ of a dollar, a goose $\frac{1}{4}$ of a dollar, and 2 chickens $\frac{1}{2}$ of a dollar: how much will the whole cost?

12. James spends $6\frac{1}{4}$ cents for candy, $12\frac{1}{2}$ cents for a top, and $5\frac{1}{5}$ cents for a slate: how much does he spend in all?

13. A man travelled $2\frac{1}{5}$ miles the first hour, $3\frac{1}{4}$ miles the second, and $4\frac{1}{2}$ the third: how far did he travel in the three hours?

14. What amount was paid for 4 weeks board, the board for the first week being $5\frac{1}{2}$ dollars, for the second $3\frac{1}{6}$ dollars, for the third $4\frac{1}{3}$ dollars, and for the fourth $4\frac{5}{6}$ dollars?

15. A man paid $\frac{2}{3}$ of a dollar for a breakfast, $1\frac{6}{7}$ of a dollar for dinner, $\frac{2}{7}$ of a dollar for supper, and $\frac{3}{7}$ of a dollar for lodging: what did he pay for the days' entertainment?

16. Jane paid $3\frac{2}{5}$ cents for tape, $6\frac{2}{3}$ cents for needles, and $4\frac{11}{15}$ cents for ribbon: how much did she pay in all?

17. If 5 yards of muslin cost $\frac{7}{8}$ of a dollar, 9 pairs of stockings $2\frac{3}{4}$ dollars, and 2 pairs of boots $9\frac{1}{2}$ dol lars, what will be the whole cost?

18. A laborer sawed wood for 4 hours, and was to have 36 cents a cord. The first hour he sawed $\frac{3}{5}$ of a cord, the second $\frac{2}{3}$ of a cord, the third $\frac{1}{2}$ of a cord, and the 4th hour $\frac{1}{10}$ of a cord: how much ought he to receive?

LESSON XIII.

Subtracting Fractions.

1. What is necessary in order that one fraction may be subtracted from another?

That both fractions have the same integral and the same fractional unit.

2. From $\frac{4}{7}$ subtract $\frac{2}{7}$.

Since the integer unit is the abstract unit one, and the fractional unit in both, is $\frac{1}{7}$, the difference is found by subtracting 2 sevenths from 4 sevenths, which leaves 2 sevenths.

3. What is the difference between $\frac{6}{15}$ and $\frac{5}{15}$?
4. What is the difference between $\frac{8}{13}$ and $\frac{5}{13}$?
5. What is the difference between $\frac{14}{21}$ and $\frac{7}{21}$?
6. What is the difference between $\frac{17}{40}$ and $\frac{14}{40}$?
7. What is the difference between $\frac{19}{25}$ and $\frac{15}{25}$?
8. What is the difference between $\frac{36}{19}$ and $\frac{17}{19}$?
9. What is the difference between $\frac{42}{18}$ and $\frac{40}{18}$?
10. What is the difference between $\frac{14}{27}$ and $\frac{12}{27}$?
11. What is the difference between $\frac{29}{31}$ and $\frac{17}{31}$?
12. What is the difference between $\frac{16}{37}$ and $\frac{9}{37}$?
13. What is the difference between $\frac{17}{37}$ and $\frac{9}{37}$?
14. What is the difference between $\frac{1}{2}$ and $\frac{1}{3}$? Reduce both to the fractional unit one sixth.

One half is equal to $\frac{3}{6}$, and one third to $\frac{2}{6}$; hence, their difference is equal to $\frac{1}{6}$.

15. What is the difference between $\frac{4}{8}$ and $\frac{1}{2}$?
16. What is the difference between $\frac{3}{7}$ and $\frac{2}{12}$?
17. What is the difference between $\frac{5}{11}$ and $\frac{2}{5}$?

18. What is the difference between $\frac{16}{17}$ and $\frac{1}{2}$?
19. What is the difference between $\frac{5}{7}$ and $\frac{2}{35}$?
20. What is the difference between $\frac{11}{13}$ and $\frac{9}{26}$?
21. What is the difference between $\frac{5}{14}$ and $\frac{9}{42}$?
22. What is the difference between $\frac{11}{13}$ and $\frac{44}{52}$?
23. From $2\frac{1}{5}$ take $1\frac{1}{4}$*
24. What is the difference between $5\frac{1}{3}$ and $2\frac{3}{4}$?
25. What is the difference between $3\frac{1}{2}$ and $2\frac{1}{6}$?
26. What is the difference between $1\frac{4}{9}$ and $\frac{5}{3}$?
27. What is the difference between $2\frac{1}{3}$ and $1\frac{4}{9}$?
28. What is the difference between $1\frac{7}{8}$ and $2\frac{1}{4}$?

QUESTIONS.

1. If you give $\frac{1}{2}$ of an orange to one boy and $\frac{1}{4}$ to another, how much more do you give to one than to the other?

2. If I have $\frac{3}{4}$ of a dollar and give $\frac{1}{2}$ dollar for a knife, how much would I have left?

3. William had $\frac{5}{6}$ of a dollar and gave $\frac{4}{9}$ of a dollar to a beggar, how much had he left?

---

* ANALYSIS.—First: $2\frac{1}{5}$ are equal to $\frac{11}{5}$, equal to $\frac{44}{20}$; and $1\frac{1}{4}$ are equal to $\frac{5}{4}$ equal $\frac{25}{20}$; hence, the difference is $\frac{19}{20}$.

It is generally best to subtract the integral and fractional numbers separately: thus, in example 25, $3\frac{1}{2}$, less $2\frac{1}{6}$, we may say 3 less 2 equals 1; $\frac{1}{2}$ less $\frac{1}{6}$ equals $\frac{2}{6}$, equals $\frac{1}{3}$: hence, the true difference is $1\frac{1}{3}$.

If the fractional part of the subtrahend is of greater value than the fractional part of the minuend, take one of the integral units of the minuend and add it to the fractional part, and then subtract.

Thus, in example 24, $\frac{3}{4}$ is greater than $\frac{1}{3}$: hence, we take 1 unit from 5 which added to $\frac{1}{3}$ makes $\frac{4}{3}$; and $\frac{4}{3}$ less $\frac{3}{4}$ leaves $\frac{7}{12}$, then adding 1 to the next figure in the subtrahend, which is the same as taking 1 from the minuend, we have 3 from 5 leaves 2: therefore, the difference between $5\frac{1}{3}$ and $2\frac{3}{4}$ is $2\frac{7}{12}$.

4. B travels $\frac{4}{5}$ of a mile in the same time that C travels $\frac{2}{3}$: which travels the farthest and how much?

5. A merchant sells $\frac{1}{2}$ of a barrel of sugar from a barrel $\frac{5}{7}$ full: what part was there left?

6. A tailor cut $\frac{7}{8}$ of a yard of cloth from a piece containing $1\frac{3}{4}$ yards: how much was there left?

7. John pays $\frac{6}{9}$ of a shilling for a knife, and $\frac{1}{3}$ of a shilling for a top: for which does he pay the most? How much?

8. Four pounds of tea cost $\frac{15}{6}$ dollars, and twenty pounds of sugar $\frac{18}{10}$: which costs the most? How much?

9. A farmer buys a calf, for which he pays $\frac{7}{3}$ dollars, and a lamb, for which he pays $\frac{9}{5}$ dollars: for which does he pay the most? How much?

10. James' shoes cost $\frac{15}{9}$ dollars, and his vest $\frac{17}{10}$ dollars: what is the difference of their cost?

11. A man earned in 4 days of a week, $5\frac{3}{7}$ dollars, and paid $1\frac{1}{3}$ dollars for his board, the three other days: how much should he receive?

12. From a piece of cloth which was $12\frac{5}{8}$ yards long, $3\frac{3}{4}$ yards is cut: how many yards are left?

13. If from a box of sugar containing $18\frac{3}{8}$ pounds, $6\frac{3}{4}$ pounds are taken, how much will be left?

14. A grocer bought $16\frac{4}{9}$ bushels of beans, and after selling $5\frac{2}{3}$ bushels, how many has he left?

15. Jane is $15\frac{6}{9}$ years old, and Nancy is $9\frac{5}{7}$ years old: how many years is Jane older than Nancy?

16. A draper cuts $5\frac{6}{7}$ yards of cloth from a piece $21\frac{11}{14}$ yards long: how much is left?

17. A grocer purchases a box of eggs, for which he paid $3\frac{7}{9}$ dollars, and sold them for $5\frac{3}{4}$ dollars: how much did he make?

18. A grocer bought a pair of chickens for $\frac{8}{9}$ of a dollar, and sold them for $\frac{9}{8}$ dollars: did he make or lose, and how much?

LESSON XIV.

Multiplication of Fractions.

1. How many are 3 times $\frac{2}{3}$?

Here the fractional unit is one third, and there are two fractional units in the expression, which being taken 3 times, gives six thirds, or 2 for the product.

2. How many are 5 times $6\frac{3}{6}$?* 8 times $\frac{4}{52}$?
3. How many are 2 times $\frac{1}{8}$? 6 times $\frac{2}{7}$?
4. How many are 8 times $\frac{9}{8}$? 7 times $\frac{2}{14}$?
5. How many are 4 times $\frac{3}{6}$? 3 times $\frac{1}{9}$?
6. How many are 7 times $\frac{4}{27}$? 6 times $2\frac{1}{2}$?
7. How many are 3 times $2\frac{1}{3}$? 5 times $3\frac{1}{5}$?
8. How many are 8 times $9\frac{1}{7}$? 4 times $6\frac{2}{4}$?
9. How many are 4 times $6\frac{1}{8}$? 3 times $9\frac{1}{3}$?
10. How many are 10 times $5\frac{2}{5}$? 9 times $6\frac{4}{18}$?
11. How many are 7 times $6\frac{3}{7}$? 9 times $9\frac{2}{18}$?
12. How many are 5 times $3\frac{4}{6}$? 8 times $6\frac{3}{7}$?
13. How many are 7 times $4\frac{5}{9}$? 6 times $3\frac{5}{3}$?
14. If the denominator of a fraction be divided by a whole number, how will the fraction be affected?

It will be increased as many times as there are units in the divisor. (See Analysis, page 86.)

How then may a fraction be multiplied by a whole number.

Either by multiplying the numerator, or dividing the denominator by the multiplier.

15. What is the product of $\frac{2}{2}$ by 2? both ways.
16. What is the product of $\frac{3}{4}$ by 4? both ways.
17. What is the product of $\frac{6}{7}$ by 7? both ways.
18. What is the product of $\frac{6}{8}$ by 4? both ways.
19. What is the product of $\frac{5}{9}$ by 9? both ways.

---

* SUGGESTION.—Let the fractional and integral units be multiplied separately: thus, 5 times 3 sixths are 15 sixths, equal to 2 and a half; and 5 times 6 are 30, to which add $2\frac{1}{2}$, giving $32\frac{1}{2}$ for the product.

20. What is the product of $\frac{3}{16}$ by 8? both ways.
21. What is the product of $\frac{6}{12}$ by 3? both ways.
22. What is the product of $\frac{3}{8}$ by 8? both ways.
23. What is the product of $\frac{7}{4}$ by 4; both ways.
24. What is the product of $\frac{4}{15}$ by 5? both ways.
25. If $\frac{3}{4}$ be multiplied by 4, what is the product?
26. What is the cost of $\frac{3}{4}$ of a yard of cloth at $\frac{5}{6}$ dollars a yard? What is $\frac{3}{4}$ of $\frac{5}{6}$?* What part is it of 1?†
27. What is the cost of $\frac{5}{7}$ boxes of raisins, at $\frac{7}{8}$ dollars a box?
28. What part of 1 is $\frac{3}{7}$ of $\frac{2}{3}$? What is $\frac{1}{7}$ of $\frac{2}{3}$?
29. What is the value of $\frac{2}{5}$ of $\frac{3}{4}$? What par of 1?
30. What is the value of $2\frac{1}{2}$ of $1\frac{1}{4}$?
31. What is the value of $\frac{5}{8}$ of $\frac{3}{7}$ of $\frac{1}{2}$?
32. What is the value of $\frac{4}{9}$ of 2?
33. How many times 1 is $3\frac{1}{3}$ of $\frac{2}{5}$?
34. How many times 1 is $2\frac{3}{4}$ of $1\frac{5}{3}$?
35. How many times 1 is $3\frac{1}{2}$ of 2?
36. How many times 1 is $5\frac{1}{2}$ of 6?
37. How many times 1 is $\frac{1}{2}$ of $\frac{1}{3}$ of $\frac{6}{7}$?
38. How many times 1 is $\frac{2}{3}$ of $\frac{3}{7}$ of $\frac{7}{2}$?
39. How many times 1 is $\frac{2}{5}$ of $2\frac{1}{2}$ of 3?
40. How many times 1 is $\frac{3}{7}$ of $\frac{1}{6}$ of 7.
41. How many times 1 is $\frac{3}{4}$ of $1\frac{1}{2}$?

---

* The word or signifies multiplication.

† ANALYSIS.—If the price per yard, $\frac{5}{6}$ of a dollar, be multiplied by $\frac{3}{4}$, the number of yards, the product will be the cost.

Now, if 1 yard of cloth cost $\frac{5}{6}$ of a dollar, $\frac{1}{4}$ of a yard will cost one fourth as much, that is, $\frac{5}{24}$ of a dollar, and $\frac{3}{4}$ will cost 3 times as much as $\frac{1}{4}$, that is, $\frac{15}{24}$, or $\frac{5}{8}$ of a dollar. Hence:

The product of two fractions is found by multiplying the numerators and denominators together.

QUESTIONS.

1. What will 5 yards of cloth cost, at $\$1\frac{3}{5}$ a yard?

2. What will 4 pounds of tea cost, at $5\frac{4}{12}$ shillings a pound?

3. What will 7 dozen of apples cost, at $11\frac{5}{7}$ cents a dozen?

4. What will 9 dozen of oranges cost, at $3\frac{7}{9}$ shillings a dozen?

5. What will 5 pairs of shoes cost, at $\$2\frac{4}{5}$ a pair?

6. What will 10 hats cost, at $3\frac{7}{10}$ dollars apiece?

7. What will $4\frac{5}{7}$ yards of cloth cost, at \$7 a yard?

8. What will be the cost of $9\frac{4}{5}$ yards, at \$5 a yard?

9. What will 12 pounds of coffee cost, at $11\frac{5}{6}$ cents a pound?

10. What will 9 sheep cost, at $\$2\frac{5}{9}$ a head?

11. What will 8 calves cost, at $\$4\frac{1}{4}$ a head?

12. What will 11 quills cost, at $1\frac{1}{4}$ cents apiece?

13. What will 10 yards of carpet cost, at $\$2\frac{4}{5}$ a yard?

14. What will 9 pairs of gloves cost, at $\$\frac{3}{4}$ a pair?

15. What will 6 pairs of fowls cost, at $\frac{5}{8}$ of a dollar a pair?

16. What will 9 pairs of boots cost, at $5\frac{1}{4}$ dollars a pair?

17. What will be the cost of 6 yards of cloth, at $4\frac{5}{6}$ dollars a yard?

18. What is the product of $2\frac{1}{2}$ of $\frac{2}{5}$ by $\frac{1}{8}$?

19. Bought $\frac{2}{3}$ of $1\frac{3}{4}$ yards of cloth at $\frac{3}{6}$ of $\frac{1}{2}$ of 4 dollars a yard: what did it come to?

20. If raisins are $2\frac{1}{4}$ dollars a box, what will be the cost of 16 boxes?

21. What will be the cost of 11 hats, if 1 hat cost $3\frac{8}{5}$ dollars?

22. If 1 pair of shoes cost $\$2\frac{5}{7}$ dollars, what will be the cost of 7 pairs?

23. What will 12 penknives cost, at $\$1\frac{1}{3}$ apiece?

24. What will 8 lb. of tea cost, at $\$1\frac{5}{11}$ a pound?

25. What will $12\frac{7}{8}$ lb. of butter cost, at 12 cents a pound?

26. What will 20 bushels of wheat cost, at $\$\frac{7}{8}$ a bushel?

27. What will 7 chickens cost, at $\$\frac{5}{8}$ apiece?

28. What will 9 turkeys cost, at $\$1\frac{1}{8}$ apiece?

29. What will 12 geese cost, at $\$\frac{3}{4}$ apiece?

30. If 1 man can earn $1\frac{1}{2}$ dollars a day, how much can 5 men earn?

31. If 1 yard of cloth cost $2\frac{3}{5}$ dollars, how much will $2\frac{1}{4}$ yards cost?

32. A grocer bought $\frac{3}{4}$ of a box of raisins for which he paid $2\frac{1}{4}$ dollars a box: what did they cost him?

33. If $\frac{3}{4}$ of a sheep is worth 3 dollars, what will be the cost of 2 sheep?

34. What will $\frac{4}{7}$ of a yard of cloth cost, at 7 dollars a yard?

35. If it cost $2\frac{2}{3}$ dollars to build 1 rod of wall, how much will it cost to build $3\frac{1}{2}$ rods?

36. James gave John $1\frac{1}{2}$ apples and had 3 times as many left, how many had he at first?

37. If a yard of calico cost $\frac{3}{4}$ of a shilling, what will $2\frac{1}{2}$ yards cost?

38. What will 12 yards of cloth cost, at $2\frac{3}{4}$ dollars a yard?

39. What will be the cost of 6 turkeys at $1\frac{1}{8}$ dollars apiece?

40. If 2 yards of cloth cost $8\frac{2}{3}$ dollars, what will 12 yards cost?

41. James gave $8\frac{1}{2}$ dimes for 10 sheets of drawing-paper: how much was that a sheet?

42. What will be the cost of a bushel of wheat, if 8 bushels cost $12\frac{3}{4}$ dollars?

LESSON XV.

Dividing Fractions.

1. What is the quotient of $\frac{6}{5}$ divided by 2?*
2. What is the quotient of $\frac{2}{2}$ divided by 2? both ways.
3. What is the quotient of $\frac{3}{6}$ divided by 3? both ways.
4. What is the quotient of $\frac{6}{8}$ divided by 3? both ways.
5. What is the quotient of $\frac{10}{9}$ divided by 5? both ways.
6. What is the quotient of $\frac{14}{6}$ divided by 7? both ways.
7. What is the quotient of $\frac{8}{12}$ divided by 4? both ways.
8. What is the quotient of $\frac{18}{7}$ divided by 9? both ways.
9. What is the quotient of $\frac{12}{15}$ divided by 6? both ways.
10. What is the quotient of $\frac{5}{8}$ divided by 5?
11. What is the quotient of $\frac{3}{7}$ divided by 3?
12. What is the quotient of $\frac{24}{8}$ divided by 4?
13. What is the quotient of $\frac{9}{5}$ divided by 9?
14. What is the quotient of $\frac{6}{8}$ divided by 2?
15. What is the quotient of $\frac{16}{12}$ divided by 8?

---

* ANALYSIS.—Here the fractional unit is $\frac{1}{5}$, and there are 6 taken. Now, if we divide the 6 fractional units by 2, the quotient will be 3; that is, $\frac{3}{5}$.

Again, if we multiply the denominator by 2, we shall have $\frac{6}{10}$, in which the fractional unit is $\frac{1}{10}$, (which is the half of $\frac{1}{5}$,) and since the number taken is the same, it follows that $\frac{6}{10}$ is on half of $\frac{6}{5}$. Hence:

A fraction may be divided by a whole number, either by dividing the numerator or multiplying the denominator by the divisor.

16. What is the quotient of $\frac{4}{12}$ divided by 4?
17. What is the quotient of $\frac{3}{12}$ divided by 3?
18. What is the quotient of $\frac{9}{12}$ divided by 3?
19. What is the quotient of $\frac{24}{8}$ divided by 12?
20. What is the quotient of $3\frac{1}{2}$ divided by 6?
21. What is the quotient of $5\frac{3}{4}$ divided by 7?
22. What is the quotient of $9\frac{4}{5}$ divided by 5?
23. What is the quotient of $12\frac{4}{3}$ divided by 8?
24. What is the quotient of $9\frac{5}{8}$ divided by 9?
25. What is the quotient of $5\frac{6}{7}$ divided by 7?
26. What is the quotient of $8\frac{3}{5}$ divided by 43?
27. What is the quotient of $9\frac{4}{5}$ divided by 7?
28. What is the quotient of $6\frac{3}{7}$ divided by 9?
29. What is the quotient of $1\frac{11}{12}$ divided by 23?
30. What is the quotient of $2\frac{4}{12}$ divided by 8?
31. What is the quotient of $\frac{1}{2}$ divided by $\frac{2}{3}$?*
32. What is the quotient of $\frac{3}{4}$ divided by $\frac{3}{5}$?
33. What is the quotient of $2\frac{1}{5}$ divided by $1\frac{1}{10}$?
34. What is the quotient of $6\frac{4}{10}$ divided by $3\frac{2}{5}$?
35. What is the quotient of $8\frac{3}{7}$ divided by $3\frac{2}{7}$?
36. What is the quotient of $3\frac{5}{8}$ divided by $2\frac{3}{8}$?
37. What is the quotient of $6\frac{4}{7}$ divided by $\frac{2}{7}$?
38. What is the quotient of $5\frac{4}{9}$ divided by 7?
39. What is the quotient of $6\frac{1}{9}$ divided by $2\frac{1}{3}$?
40. What is the quotient of $7\frac{4}{5}$ divided by $2\frac{6}{10}$?
41. What is the quotient of $8\frac{3}{7}$ divided by $6\frac{1}{7}$?
42. What is the quotient of $5\frac{1}{4}$ divided by $5\frac{1}{3}$?
43. What is the quotient of $8\frac{1}{3}$ divided by $2\frac{1}{2}$?
44. What is the quotient of $3\frac{4}{5}$ divided by $6\frac{2}{10}$?

---

* SUGGESTION.—Reduce both to sixths: $\frac{1}{2}$ is equal to $\frac{3}{6}$, and $\frac{2}{3}$ are equal to $\frac{4}{6}$. Now, since the fractional unit is the same in both, the quotient will be found by dividing the numerators: Hence, to divide one fraction by another,

Reduce them to the same fractional unit, and then divide the numerator of the dividend by the numerator of the divisor.

45. What is the quotient of $6\frac{2}{3}$ divided by $3\frac{1}{2}$?
46. What is the quotient of $8\frac{2}{7}$ divided by 2 ?
47. What is the quotient of $4\frac{3}{4}$ divided by $6\frac{1}{2}$?
48. What is the quotient of $2\frac{4}{9}$ divided by $3\frac{1}{3}$?
49. What is the quotient of $4\frac{3}{5}$ divided by $2\frac{1}{10}$?
50. What is the quotient of $8\frac{3}{4}$ divided by $6\frac{1}{2}$?
51. What is the quotient of $3\frac{7}{9}$ divided by $2\frac{5}{3}$?
52. What is the quotient of $6\frac{3}{10}$ divided by $2\frac{2}{5}$?
53. What is the quotient of $3\frac{4}{5}$ divided by $8\frac{4}{5}$?
54. What is the quotient of $9\frac{1}{9}$ divided by $3\frac{4}{3}$?
55. How many times is $\frac{2}{5}$ contained in $\frac{3}{4}$?*
56. How many times is $\frac{4}{9}$ contained in 2 ? $\frac{1}{9}$ is contained in 1 how many times ? In 2 how many times ? How many times are $\frac{4}{9}$ contained ?
57. How many times is $\frac{3}{4}$ contained in $\frac{7}{8}$?
58. How many times is $2\frac{1}{5}$ contained in 3 ?
59. How many times is $\frac{4}{7}$ contained in $\frac{5}{4}$?
60. How many times is $\frac{3}{8}$ contained in $2\frac{1}{2}$?
61. How many times is $\frac{3}{7}$ contained in $\frac{5}{9}$?
62. How many times is $2\frac{1}{3}$ contained in $\frac{4}{5}$?
63. How many times is $\frac{5}{7}$ contained in $\frac{5}{6}$?

QUESTIONS.

1. If 3 yards of cloth cost $\$10\frac{1}{2}$, how much does it cost a yard ?
2. If six pounds of tea cost $\$4\frac{4}{8}$, what does it cost a pound ?
3. If John gives $9\frac{3}{7}$ cents for 7 tops, how much do they cost him apiece ?
4. If 7 pounds of sugar cost $\frac{3}{8}$ of a dollar, how much is it a pound ?

---

* ANALYSIS.—One is contained in $\frac{3}{4}$, $\frac{3}{4}$ times. But $\frac{1}{5}$ is contained in $\frac{3}{4}$, 5 times as many times as 1 ; that is, $\frac{15}{4}$ times. But 2 fifths is contained half as many times as $\frac{1}{5}$; that is, $\frac{15}{8}$ times : Hence, to find the quotient of one fraction divided by another,

Invert the terms of the divisor and multiply.

5. If 4 pounds of coffee cost \$$1\frac{7}{9}$, how much does it cost a pound?

6. If 7 oranges cost $9\frac{1}{3}$ cents, how much do they cost apiece?

7. If three and three fourths yards of cloth cost \$$11\frac{1}{4}$, how much does it cost a yard?

8. If James can walk 14 miles in $\frac{7}{9}$ of a day, how far can he walk in one day?

9. If John can buy 9 lemons for $10\frac{1}{2}$ cents, how much do they cost him apiece?

10. If 9 eggs cost $10\frac{4}{5}$ cents, how much do they cost apiece?

11. If $7\frac{1}{2}$ bunches of grapes are worth $22\frac{1}{2}$ shillings, how much are they a bunch?

12. If $5\frac{1}{5}$ bushels of potatoes cost \$$2\frac{3}{5}$, how much do they cost a bushel?

13. If nine baskets of peaches cost \$$12\frac{3}{5}$, how much are they a basket?

14. If 8 lambs cost \$$12\frac{4}{5}$, how much do they cost apiece?

15. If $9\frac{3}{7}$ pounds of cheese cost \$$\frac{3}{7}$, how much does it cost a pound?

16. If 4 barrels of flour cost 24 dollars, what will $1\frac{3}{6}$ cost?

17. James has 3 oranges and 5 playmates: he wishes to give $\frac{1}{3}$ of an orange to each, how must he divide the oranges, and how many oranges will he have left?

18. If one man consumes $1\frac{1}{5}$ pounds of meat in a day, how many men would $8\frac{2}{5}$ pounds supply?

19. A man distributed $17\frac{3}{5}$ pounds of bread among 8 persons, how much does he give to each?

20. If 12 horses consume $28\frac{4}{5}$ tons of hay in a winter, how much is consumed by each horse?

21. If $\frac{3}{5}$ of a barrel of flour will last a family 30 days, how long will 2 barrels last them?

22. A farmer has a field containing $8\frac{3}{4}$ acres; if a man can mow $1\frac{3}{4}$ acres in a day, how many men will it take to mow the field in 1 day?

23. If 9 bushels of apples cost $4\frac{1}{2}$ dollars, how much will 17 bushels cost?

24. If 7 pounds of butter cost $10\frac{1}{2}$ shillings, how much is that a pound?

25. If 3 pounds of butter cost $7\frac{1}{2}$ shillings, how much will 12 pounds cost?

26. If 5 yards of cloth cost $11\frac{1}{4}$ dollars, what will 8 yards cost?

27. If 8 yards of cloth cost $42\frac{1}{2}$ dollars, how much will 4 yards cost?

28. If $1\frac{2}{3}$ dollars will buy 2 yards of cloth, how many yards will 6 dollars buy?

29. How many times is $2\frac{1}{2}$ contained in $12\frac{1}{2}$?

30. How many pounds of tea can be purchased for $6\frac{4}{5}$ dollars, if it cost $\frac{4}{5}$ dollars a pound?

31. If a turkey costs $1\frac{3}{5}$ dollars, how many can be bought for $12\frac{4}{5}$ dollars?

32. If calico is worth $\frac{1}{8}$ of a dollar a yard, and muslin $\frac{1}{7}$ of a dollar a yard, how much calico must be given for 9 yards of muslin? What is the cost of 9 yards of muslin?

33. A tailor bought a piece of cloth containing $12\frac{2}{3}$ yards, for which he paid 38 dollars: what did it cost him a yard?

34. If you give $15\frac{2}{3}$ dollars for a cow, and sell her for $\frac{2}{5}$ more than she cost, how much more do you receive for her than you gave?

35. If 6 men can do a piece of work in $15\frac{3}{5}$ days, how long will it take 1 man to do it? 2 men? 3 men?

36. A man divided $2\frac{4}{5}$ of a dollar among his children, giving $\frac{7}{10}$ of a dollar to each: how many children were there.

SECTION SEVENTH.

LESSON I.

Comparison of Numbers.

1. What part of 2 is 1 ? How many times does 1 ontain 2 ?*

2. What part of 3 is 2 ? What is the ratio of to 2 ?

3. What part of 6 is 3 ? How many times does 3 contain 6 ?

4. What part of 12 is 4 ? What is the ratio of 12 to 4 ?

5. What part of 4 is 1 ? What is the ratio ?

6. What part of 3 is 2 ? What is the ratio ?

7. What part of 8 is 5 ? What is the ratio ?

8. What part of 20 is 10 ? What is the ratio ?

9. What part of 30 is 20 ? What is to ratio ?

10. What part of 50 is 10 ? What is the ratio ?

11. What part of $\frac{3}{4}$ is $\frac{1}{4}$? What is the ratio ?

12. What part of $\frac{4}{9}$ is $\frac{2}{8}$? What is the ratio ?

13. What part of $\frac{4}{5}$ is $\frac{3}{6}$? What is the ratio ?

---

* SUGGESTIONS.—To find what part any number is of a number less than itself, we divide the less by the greater, and the quotient shows the part. This quotient is called the *ratio*. Thus, what part of 2 is 1 ; it is 1 divided by 2, which is $\frac{1}{2}$.

If we wish to find how many times one number is greater than another, we divide the greater by the less, and the quotient shows the number of times. This quotient is also called the *ratio* of the two numbers. Thus, how many times is 12 greater than 2: it is 12 divided by 2 times greater, which is 6 times.

But the general question, "How many times," has no reference to the relative value of the numbers. Thus: what part of 2 is 1 ? Or, how many times is 2 contained in 1 ? One half, or one half times ? How many times is 2 contained in 4 ? Twice, or 2 times.

14. What part of $\frac{3}{7}$ is $\frac{2}{8}$? What is the ratio?
15. What part of $\frac{8}{9}$ is $\frac{4}{10}$? What is the ratio?
16. What part of $\frac{7}{3}$ is $\frac{3}{5}$? What is the ratio?
17. What part of $3\frac{1}{2}$ is $3\frac{3}{4}$? What is the ratio?
18. What part of $4\frac{3}{5}$ is $3\frac{1}{4}$? What is the ratio?
19. What part of $3\frac{1}{2}$ is $2\frac{4}{8}$? What is the ratio?
20. What part of $6\frac{1}{4}$ is $5\frac{1}{3}$? What is the ratio?
21. What part of 8 is $\frac{2}{3}$ of $2\frac{1}{2}$? What is the ratio?
22. What part of 10 is $\frac{3}{4}$ of $4\frac{2}{3}$? What is the ratio?
23. What part of 15 is 4?
24. What part of 45 is 12?
25. What part of $\frac{7}{8}$ is $\frac{5}{6}$?
26. What is the ratio of 12 to 7?
27. What is the ratio of $\frac{11}{12}$ to $\frac{1}{2}$?
28. What is the ratio of $3\frac{5}{6}$ to $1\frac{4}{12}$?
29. What is the ratio of 9 to $\frac{4}{5}$ of $7\frac{1}{2}$?
30. What part of $7\frac{1}{2}$ is $1\frac{1}{3}$ of $3\frac{1}{4}$?

QUESTIONS.

1. What part of 30 dollars is 5 dollars? 6 dollars? 10 dollars? 2 dollars?

2. What part of 56 bushels is 8 bushels? 7 bushels? 4 bushels? 28 bushels?

3. What part 25 yards is $\frac{3}{8}$ of 16 yards?

4. A tailor has 12 yards of cloth; he cuts off 3 yards for a coat, $1\frac{1}{4}$ yards for a pair of pantaloons, $1\frac{3}{4}$ of a yard for a vest: what part of the cloth did he use? What part was left?

5. A farmer has 60 bushels of wheat in his barn; he sells 20 bushels to one man, 15 to another, and 10 to another: what part of the wheat has he left?

6. Mr. Dix buys $\frac{2}{3}$ of $\frac{1}{2}$ of $\frac{3}{4}$ of Mr. Jones' farm, which contains 100 acres: what part of the farm does buy? How many acres are there left?

7. James gave 5 dollars for his hat, and his father gave 25 dollars for a coat: what part of the cost of the coat was the cost of the hat?

8. What part of a barrel of flour is $\frac{1}{2}$ of $\frac{3}{5}$ of it? What part of the whole barrel is the remainder? What will each part cost if flour is 8 dollars a barrel?

9. William having a pine-apple said he would give $\frac{2}{3}$ of $\frac{5}{7}$ of it to the one that could tell how much that would be: how much did the individual receive? What part of the pine-apple was left?

10. A merchant had $4\frac{3}{4}$ barrels of flour and sold $\frac{2}{3}$ of it: how much had he left?

11. A lady paid $\frac{3}{5}$ of $5\frac{1}{3}$ dollars for a pencil and $1\frac{3}{5}$ dollars for a ring: how many times as much did she pay for the pencil as for the ring?

12. One man can build $5\frac{5}{9}$ rods of wall in one day, and another can build $3\frac{1}{3}$ rods: what part is the second of the first?

13. Albert is $9\frac{3}{7}$ years of age, John is $\frac{5}{6}$ of Albert's age: how old is John?

14. James bought a kite for $16\frac{3}{4}$ cents, which was $2\frac{1}{2}$ times as much as he paid for his top: how much did he pay for his top?

15. A farmer bought a calf for $4\frac{3}{8}$ dollars, and a pig for $1\frac{4}{9}$ of that sum: what did he pay for the pig?

16. A box of raisins cost $3\frac{1}{3}$ dollars, at the rate of $\frac{1}{4}$ of a dollar a pound: how many pounds were there in the box?

17. If 14 oranges are worth as much as $3\frac{1}{2}$ pine-apples, how many oranges is one pine-apple worth?

18. Mary gave $\frac{7}{8}$ of $3\frac{1}{3}$ dollars for a bonnet, which was $\frac{1}{3}$ of what Jane paid for hers: how much did Jane pay?

19. If a father can do $1\frac{1}{8}$ times more work in a day than his son, how many days' work of the father will be equal to 18 of the son's?

20. A gentleman gave 200 dollars to his three sons; to the first he gave $\frac{2}{5}$ of it, to the second $\frac{5}{8}$ of 2 times as much, and the rest to the third: how much had each?

21. If a boy can run 80 rods in eight minutes, what part of a mile can he run in 16 minutes?

22. James bought $\frac{1}{2}$ pound of candy, and gave $\frac{4}{5}$ of it to Mary: what part of a pound did he give her?

23. A vessel sails 150 miles one day, and is retarded $\frac{1}{4}$ of $\frac{4}{5}$ as much on the second: what part of 150 miles did she sail the second day?

LESSON II.

Comparison of Numbers.

1. Two is one half of what number?*
2. Four is one third of what number?
3. Five is one fifth of what number?
4. Two and one half is one-fourth of what number?
5. One and one half is one sixth of what number?
6. One and 2 thirds is one-third of what number?†
7. Three and one third is one third of what number?
8. Ten and five sixth is one sixth of what number?
9. Twelve is one fifth of what number?
10. Eight and three fourths is one fourth of what number?
11. Nine and one sixth is one sixth of what number?
12. Five and one tenth is one tenth of what number?

* ANALYSIS.—Two is one half of 2 times 2, which are 4 therefore, 2 is one half of 4.

† ANALYSIS.—One and 2 thirds is one third of 3 times one and 2 thirds, which are 5: therefore, one and two thirds is one third of 5.

13. Five and $\frac{3}{5}$ is one fifth of what number?

14. Six and three fourths is one eighth of what number?

15. Twelve and one sixth is one sixth of what number?

16. Two and three fourths is one twelfth of what number?

17. Three and six ninths is one ninth of what number?

18. Six and $\frac{2}{3}$ is one twelfth of what number?

19. Eight is 2 thirds of what number?*

20. 12 is 3 fourths of what number?

21. 16 is 4 fifths of what number?

22. 30 is 6 sevenths of what number?

23. 32 is 8 ninths of what number?

24. 15 is 5 ninths of what number?

25. 12 is 4 sevenths of what number?

26. 22 is 11 twelfths of what number?

27. 27 is 9 tenths of what number?

28. 21 is 7 ninths of what number?

29. 15 is 3 sevenths of what number?

30. 24 is 8 elevenths of what number?

31. 16 is 5 times what number?†

32. 48 is 6 times what number?

33. 39 is 3 times what number?

34. 56 is 9 times what number?

35. 21 is 5 times what number?

36. 75 is 8 times what number?

37. 54 is 8 times what number?

38. 95 is 9 times what number?

---

* Analysis.—Since 8 is 2 thirds of some number, one half of 8, which is 4, is one third of the same number; but 4 is one third of 3 times 4, which are 12; therefore, 8 is 2 thirds of 12.

† Analysis.—16 is 5 times $\frac{1}{5}$ of 16, which is $3\frac{1}{5}$: therefore, 16 is 5 times $3\frac{1}{5}$.

QUESTIONS.

1. Two thirds of nine is one half of what number?

2. Two sevenths of fourteen is one third of what number?

3. Three twelfths of thirty-six is one eighth of what number?

4. James gave nine cents for a slate, which was three fourths of his money: how much had he?

5. A man bought a cow, for which he paid $30, which was three fifths of his money: how much had he?

6. A lady was married at twenty years of age, which was the half of eight fifths of the age of her husband: how old was the husband?

7. In a pasture are 45 sheep, which is three fourth the number of cows in the same pasture: how many cows are there?

8. John gave 36 cents for a knife, which was six sevenths of what he gave for a sled: how much did he give for his sled?

9. If a man can make six and two ninths rods of fence in one day, how much can he make in 12 days?

10. Two men agreed to build a wall; one man built 16 rods, which was four fifths of what the other built: how much did the second build, and what was the whole length of the wall?

11. A man willed one half of his property to his wife, one third of the remainder to his daughter Mary, and one sixth to his son James: how much of it was left for his only remaining son John?

12. John gave one and 3 fourth cents for a peach which was one third of what he gave for an orange what did the orange cost him?

13. Charles gave ten and five sixths cents for his kite, which was five times what he paid for his top: how much did he pay for his top? How much for both?

14. William gave eight and five ninths cents for a pencil, which was one fourth the cost of his penknife: what did his penknife cost him?

15. A farmer paid four and three fourths dollars for a calf, and one fifth as much for a pig: what did the pig cost him?

16. A farmer bought a calf for three and one seventh dollars, which was one seventh of what he paid for a cow: what did the cow cost him?

17. A man bought a pair of boots for six and $\frac{4}{7}$ dollars, and a coat which cost him three and a half times as much: what did the coat cost him?

18. If 3 sevenths of a barrel of flour cost 6 dollars, what will 5 sevenths cost? What will the whole barrel cost?

19. A grocer bought 5 boxes of raisins for two and three fifths dollars a box, and a barrel of sugar, which cost one half of two thirds as much as the raisins: how much more did the sugar cost him than the raisins?

LESSON III.

Comparison of Numbers.

1. 25 is 5 eighths of how many times 7?*
2. 63 is 7 ninths of how many times 8?
3. 36 is 4 sevenths of how many times 6?
4. 45 is 5 sixths of how many times 5?
5. 84 is 7 eighths of how many times 9?
6. 29 is 3 ninths of how many times 10?
7. 42 is 7 thirds of how many times 5?
8. 64 is 8 fifths of how many times 3?
9. 32 is 3 eighths of how many times 11?

---

* ANALYSIS.—If 25 is 5 eighths of some number, 1 eighth is 1 fifth of 25, which is 5. If 5 is 1 eighth, 8 eighths are 8 times 5, which are 40. As many as 7 is contained times in 40: 7 is contained in 40, 5 and 5 sevenths times; therefore, 25 is 5 eighths of 5 times 7 and 5 sevenths of 7

10. 75 is 5 eighths of how many times 6?
11. 21 is 7 fifths of how many times 4?
12. 40 is 4 sevenths of how many times 3?

13. 6 sevenths of 21 is 3 fifths of what number?*
14. 8 ninths of 27 is 4 sevenths of what number?
15. 3 fourths of 40 is 6 tenths of what number?
16. 5 eighths of 64 is 8 ninths of what number?
17. 2 thirds of 42 is 5 twelfths of what number?
18. 7 tenths of 80 is 8 ninths of what number?
19. 4 fifths of 60 is 6 sevenths of what number?
20. 5 sixths of 48 is 7 eighths of what number?
21. 3 sevenths of 21 is 2 thirds of what number?
22. 4 ninths of 81 is 5 eighths of what number?
23. 7 fifths of 45 is 9 sevenths of what number?
24. 5 ninths of 36 is 4 fifths of how many times 5?
25. 4 sevenths of 56 is 8 ninths of how many times 7?
26. 3 fourths of 24 is 6 sevenths of how many times 4?
27. 5 eighths of 48 is 5 ninths of how many times 9?
28. 4 thirds of 30 is 8 elevenths of how many times 8?
29. 5 tenths of 72 is 4 ninths of how many times 6?
30. 11 twelfths of 84 is 7 ninths of how many times 10?
31. 6 fifths of 50 is 4 thirds of how many times 12?

---

* ANALYSIS.—6 sevenths of 21 is 6 times 1 seventh of 21: 1 seventh of 21 is 3, and 6 sevenths is 6 times 3, which are 18. If 18 is 3 fifths of some number, 1 third of 18 which is 6 is 1 fifth of the same number; 6 is 1 fifth of 5 times 6 which is 30: therefore, 6 sevenths of 21 is 3 fifths of 30.

VERIFICATION.—One fifth of 30 is 6, and 3 fifths of 30 are 18 but 6 sevenths of 21 are 18: therefore, 30 is the number sought.

NOTE.—With the correct answer all the questions may be reversed and similarly analyzed: thus, 30 is 5 thirds of 6 sevenths of what number? Again, 30 is 5 thirds of how many sevenths of 21?

32. 9 elevenths of 88 is 8 fifths of how many times 6?

33. 2 thirds of 75 is 5 sixths of how many times 9?

34. 5 halves of 24 is 3 fourths of how many times 5?

35. 5 ninths of 72 is 4 thirds of what number?

36. 6 sevenths of 56 is 5 ninths of what number?

37. 28 is 1 fifth of how many times 9?

38. 42 is 7 twelfths of how many times 12?

39. 5 ninths of 108 is 10 elevenths of how many times 10?

40. 6 fifths of 35 is 7 twelfths of how many times 5?

41. 81 is 9 fourths of how many times 8?

42. 7 thirds of 36 is 9 tenths of what number?

43. 7 twelfths of 96 is 8 ninths of how many times 5?

QUESTIONS.

1. A boy gave 5 apples to one of his companions, which was one third of all he had: how many had he?

2. A man bought a watch for 24 dollars, and sold it for four fifths of what it cost him: what did he receive for it, and how much did he lose by the bargain?

3. Charles gave away 12 apples to 3 of his companions, which was $\frac{2}{3}$ of $\frac{3}{5}$ of all he had: how many had he, and how many did he give each?

4. If 3 fourths of a hundred weight of sugar cost 12 dollars, what will a hundred weight cost? How many barrels of cider at 2 dollars a barrel will pay for it?

5. A man sold a horse for 60 dollars, which was 5 sevenths of what he cost him: how much did he cost him, and how much did he lose? When he bought him he paid in cloth at 6 dollars a yard: how many yards of cloth did he give?

6. A pole is $\frac{2}{5}$ in the water, $\frac{1}{4}$ in the mud, and 14 feet out of the water: how long is the pole?

7. John's age is $\frac{2}{3}$ William's, and the sum of their ages diminished by 5 is equal to 70: what is the age of each?

8. Mr. Wilson gave 200 dollars to his neice, which was $\frac{1}{4}$ of $\frac{4}{5}$ of his property, and the remainder equally to his 4 sons: how much did each receive?

9. How many yards of cloth, at 4 dollars a yard, must be given for a hogshead of sugar, if four sevenths of it cost 48 dollars?

10. There is a pole standing $\frac{3}{7}$ in the water, $\frac{1}{2}$ of the remainder in the mud, and 6 feet above the water: how long is the pole?

11. A staff 5 feet long casts a shadow of 3 feet: what is the length of a pole that casts a shadow of 24 feet the same time of day?*

12. A man can build 56 rods of wall in a certain time, another man can build 8 rods while the first builds 7: how much would the second build in the same time?

13. Two boys counting their marbles, one said he had sixteen. The other said, 3 eighths of yours is exactly 2 sevenths of mine: now if you will tell me how many I have, I will give you the difference between yours and mine: how many had he?

14. A man being asked how many sheep he had, said he had them in three pastures: in the first he had 42, which was 7 eighths of what he had in the second; and that 5 thirds of what he had in the second was just 4 times what he had in the third: how many had he in each field?

15. A gambler lost 3 fourths of his money in play; he then won 30 dollars, which was 5 sixths of what he lost: how much money had he when he began to play?

---

* ANALYSIS.—Since the shadow of the staff is 3 fifths the length of the staff, the shadow of the pole must be 3 fifths the length of the pole; then if 24 is 3 fifths of some number &c.

16. A tailor cut off 3 fifths of a piece of cloth, he then cut off 4 yards, which was one third of the remainder : how many yards were there in the piece?

17. A gentleman left to his eldest son 300 dollars, which was 3 fourths of what the second son had, and twice the second son's share was just four times what the third son received : how much was the fathe worth?

18. James being asked how many credit marks he had, said : if 1 third of the number be taken from 1 half of the number, the remainder would be $2\frac{1}{2}$ times 4 : how many credits had he?

19. Three fourths of 40 is 5 sevenths as many dollars as Mr. C. paid for his horse : what was the cost of the horse?

20. A person being asked his age said, that 3 fourths of 80 was 6 sevenths of 5 times his age : what was his age?

21. Bought 45 yards of cloth and sold 4 ninths of it for 20 dollars, which was 5 sixths of what the whole cost : what would be the gain on the whole, at the same rate?

22. A merchant bought 6 barrels of flour at 54 dollars, which was 9 eighths of what it cost him : what did it cost him a barrel?

LESSON IV.

1. 3 sevenths of 56 is 8 ninths of 3 times what number?*

* ANALYSIS.—Three sevenths of 56 is 3 times 1 seventh of 56. One seventh of 56 is 8, and 3 sevenths of 56 is 3 times 8 which are 24. Since 24 is 8 ninths of some number, 1 eighth of 24 which is 3 is 1 ninth of the same number: 3 is 1 ninth of 9 times 3 which is 27. Now, 27 is 3 times 1 third of 27 which is 9; therefore, 3 sevenths of 56 is 8 ninths of 3 times 9, or 27.

2. 5 sixths of 54 is 5 eighths of nine times what number?

3. 4 fifths of 30 is 3 fourths of 8 times what number?

4. 2 ninths of 81 is 3 elevenths of 4 times what number?

5. 5 eighths of 56 is 7 ninths of 6 times what number?

6. 4 thirds of 36 is 6 fifths of 10 times what number?

7. 9 tenths of 90 is 6 fourths of 8 times what number?

8. 6 halves of 30 is 9 tenths of 20 times what number?

9. 7 ninths of 108 is 7 twelfths of 8 times what number?

10. 5 eighths of 32 is 4 fifths of how many sixths of 18?*

11. 6 sevenths of 56 is 8 ninths of how many fourths of 24?

12. 8 fifths of 40 is 4 thirds of how many sixths of 36?

13. 4 thirds of 36 is 8 twelfths of how many fifths of 45?

14. 2 fifths of 75 is 5 sevenths of how many ninths of 54?

15. 3 halves of 40 is 6 twelfths of how many tenths of 80?

16. 7 ninths of 72 is 8 fifths of how many thirds of 24?

17. 6 elevenths of 44 is 3 tenths of how many fourths of 32?

---

* ANALYSIS.—The same as in the preceding examples until you obtain the second number, which in this example is 25. Then, 25 is how many sixths of 18? 1 sixth of 18 is 3, and 3 is contained in 25 8 and 1 third times; therefore, 5 eighths of 32 is 4 fifths of 8 and 1 third times 1 sixth of 18.

18. 5 sevenths of 77 is 11 twelfths of how many eighths of 56?

19. 10 thirds of 24 is 8 halves of how many twelfths of 36?

20. 12 fifths of 45 is 9 tenths of how many eighths of 64?

21. 5 sixths of 54 is 3 fourths of how many sevenths of 42?

22. 4 fifths of 30 is 4 sevenths of how many times 2 thirds of 21?*

23. 6 thirds of 27 is 6 ninths of how many times 1 tenth of 90?

24. 8 ninths of 63 is 7 twelfths of how many times 4 eighths of 24?

25. 7 sixths of 54 is 9 tenths of how many times 2 ninths of 45?

26. 5 halves of 24 is 5 eighths of how many times 4 sevenths of 28?

27. 9 tenths of 70 is 7 sixths of how many times 3 ninths of 27?

28. 4 thirds of 36 is 4 ninths of how many times 2 sevenths of 42?

29. 5 ninths of 72 is 4 fifths of how many times 5 twelfths of 60?

30. 7 eighths of 64 is 8 sevenths of how many times 3 eighths of 32?

31. 10 fourths of 36 is 9 thirds of how many sevenths of 63?

---

* NOTE.—In connection with the answer, a reversed statement of the examples in this lesson may be made, giving two other propositions, to be solved by a similar analysis. Thus, in Example 22, 3 being the answer we have,

First. 3 times 2 thirds of 21 is 7 fourths of 4 fifths of what number? which will give 30.

Second. 3 times 2 thirds of 21 is 7 fourths of how many fifths of 30? which will give 4 *fifths* of 30.

This will give the pupil the benefit of three examples in one

32 6 sevenths of 77 is 11 twelfths of how many fifths of 50?

33. 8 ninths of 81 is 6 fifths of 4 times what number?

34. 9 tenths of 100 is 5 halves of 8 times what number?

35. 3 sevenths of 84 is 4 ninths of how many times 4 ninths of 45.

QUESTIONS.

1. Two boys comparing their ages, one said he was fifteen years old; the other said, 4 fifths of your age is just 3 halves of my age: what was his age?

2. A farmer had a certain number of sheep which he put in two fields; in one field he had 28, and 6 sevenths of them was 4 ninths of 2 times what he had in the second: how many were there in the second field?

3. A man pays 300 dollars a year for benevolent objects: $\frac{2}{3}$ of this sum is equal to $\frac{1}{7}$ of 2 times the amount of his personal expenses: what are his personal expenses?

4. A farmer sold a number of cows and had 12 left which was $\frac{1}{3}$ of the number sold; if the number sold be divided by $\frac{3}{4}$ of $9\frac{1}{3}$, the quotient will be $\frac{1}{5}$ the number of dollars he received per head: how much did he receive apiece for his cows?

5. The insurance on a house is 600 dollars, and $\frac{1}{3}$ of that is $\frac{1}{5}$ of 4 times the value of the furniture: what is the furniture worth?

6. A man bought a horse for 100 dollars, $\frac{4}{5}$ of what the horse cost was $\frac{2}{3}$ of what he paid for a carriage how much did the carriage cost him?

7. $\frac{4}{7}$ of A's age is $\frac{4}{5}$ of B's, and 3 times B's is $\frac{5}{2}$ of C's: how old are A and B, if C is 24 years old?

8 Fort Plain is 56 miles from Albany, $\frac{5}{7}$ of this distance is $\frac{4}{5}$ times $\frac{1}{5}$ the distance from Albany to Rochester: what is the distance to Rochester?

9. The contents of a certain store cost 1,000 dollars, and $\frac{2}{5}$ the entire cost is $\frac{4}{9}$ of 3 times what the cloths cost: what was the cost of the cloths?

10. James has a certain number of marbles; John has $\frac{6}{7}$ as many less 3, and William has $\frac{4}{5}$ as many as John less 7; William has 5 marbles: how many have John and James?

11. A house is worth 600 dollars, and $\frac{5}{6}$ of its value is $\frac{1}{5}$ of $2\frac{1}{2}$ times the value of the farm on which it stands: what is the value of the farm?

12. Buffalo is 325 miles from Albany, and $\frac{3}{5}$ of this distance is $3\frac{3}{4}$ times $\frac{2}{3}$ the distance from Rochester to Buffalo: what is the distance?

13. A boy being asked his age said, that 9 years was 3 years more than $\frac{3}{4}$ times $\frac{4}{7}$ of his age: what was his age?

14. A man had $\frac{5}{6}$ of his money stolen from him; the thief was caught, but not until he had spent $\frac{2}{3}$ of it, the remainder, ($50), was given back: how much money had he at first?

15. A and B engaged in play with equal sums of money, B gained 40 dollars, which was $\frac{2}{9}$ of 3 times what he commenced with: how much had each when they began to play? How much had A left?

15. A farmer sold a horse for 96 dollars, which was $\frac{4}{5}$ times $\frac{8}{9}$ what he paid for him: how much did he pay for him?

LESSON V.

Comparison of the units of Denominate Numbers.

1. Four mills are what part of a cent?
2. Five cents are what part of a dollar?
3. Three dimes are what part of a dollar?
4. Thirty-six cents are what part of a dollar?

5. Three dollars is what part of an eagle?

6. How many cents in $\frac{3}{5}$ of a dollar?

7. How many dimes in $\frac{2}{5}$ of a dollar?

8. How many mills in $\frac{3}{10}$ of a dollar?

9. How many dollars in $\frac{3}{4}$ of an eagle?

10. What part of a pound is 1 shilling?

11. What part of a pound is 6 pence?

12. What part of a shilling is 5 pence?

13. What part of a shilling is 3 farthings?

14. What is the value, in pence, of $\frac{2}{3}$ of a shilling? Of $\frac{3}{8}$ of a shilling? Of $\frac{1}{4}$ of a pound?

15. What is the value, in shillings and pence, of $\frac{3}{5}$ of $\frac{1}{2}$ of a pound? Of $\frac{1}{3}$ of $\frac{2}{3}$ of $\frac{1}{4}$ of a pound?

16. Seven pence is what part of a pound?

17. Eleven pence is what part of a shilling? Of £?

18. What part of a pound is 6 shillings? 7 shillings? 12 shillings? 13 shillings? 14 shillings? 15 shillings?

19. What part of a pound is 3*s.* 8*d.*? 4*s.* 6*d.*? 2*s.* 7*d.*? 5*s.* 9*d.*? 6*s.* 8*d.*? 4*s.* 9*d.*?

20. What part of a shilling is $8\frac{1}{2}$*d.*? $6\frac{3}{4}$*d.*? $9\frac{1}{4}$*d.*?

21. What part of a pound is 2*oz.*? 4*oz.*? 6*oz.*? 9*oz.*? 12*oz.*?

22. What part of a quarter is 8*lb.*? What part of a quarter is 12*lb.*? 14*lb.*? 18*lb.*? 20*lb.*?

23. What part of 1*cwt.* is 3 quarters? 2 quarters? 1 quarter?

24. What part of 1*cwt.* is 15*lb.*? 27*lb.*? 95*lb.*? 75*lb.*? 68*lb.*?

25. What part of a ton is 8*cwt.*? 12*cwt.*? 14*cwt.*? 16*cwt.*? 19*cwt.*?

26. What part of a ton is 65*lb.*? What part of a ton is 95*lb.*? 350*lb.*? 1000*lb.*?

27. What part of a foot is 5 inches? 7 inches? 9 inches? 4 inches? 10 inches? 11 inches?

28. What part of a yard is 1 foot? What part is 1 foot of 2 yards? What part of 5 yards is 1 foot?

29. What part of a yard is 1 foot? What part of a yard is 2 feet?

30. What part of a furlong is 1 rod? 3 rods? 5 rods? 8 rods? 10 rods? 12 rods? 15 rods?

31. What part of a mile is 7 furlongs? 5 furlongs? 4 furlongs? 3 furlongs? 2 furlongs? 6 furlongs?

32. What part of a mile is 20 rods? 30 rods? 40 rods? 50 rods? 160 rods?

33. What part of a square foot is 12 square inches? 48 square inches? 100 square inches?

34. What part of a square yard is 2 square feet? 7 square feet? 8 square feet? 5 square feet?

35. What part of a square yard is 3 square feet? 6 square feet? 1 square foot? 4 square feet?

36. What part of a rood is 4 square rods? 8 square rods? 9 square rods? 7 square rods?

37. What part of an acre is 4 square rods? 10 square rods? 40 square rods? 100 square rods? What part of an acre is 3 roods?

38. What part of a quart is 1 pint? 2 pints is what part of 8 quarts? 3 pints is what part of 5 quarts?

39. In wine measure, what part of a quart is 1 pint? What part of a quart is 2 gills? 3 gills? 5 gills?

40. What part of a gallon is 3 quarts? 3 pints? 3 gills? 2 quarts? 5 pints? 5 gills?

41. What part of a hogshead is 1 gallon? 2 gallons? 8 gallons? 9 gallons?

42. One pint in dry measure is what part of a quart? What part of a peck? What part of a bushel?

43. Three pecks is what part of a bushel? What part of a chaldron?

44. Five minutes is what part of an hour?

8 seconds is what part of a minute? What part of an hour?

45. Three hours is what part of a day? What part of a week?

46. Four days is what part of a week? What part of a month? What part of a year?

47. Five years is what part of a century? What part of a century is 30 years? 40 years? 6 years?

48. What part of 3 days is 5 hours? 6 hours? 7 hours? 9 hours? 10 hours?

49. What part of 7 months is 9 weeks? What part is 8 weeks? 6 weeks?

50. Four minutes is what part of a day? Of an hour? Of a week? Of a month?

QUESTIONS.

1. What will $3\frac{3}{4}$ yards of cloth cost at 4 dollars a yard?

2. If 4 bushels of wheat be divided equally among 5 men, how much will each receive?

3. What will $5\frac{1}{2}$ bushels of wheat cost at $2\frac{1}{2}$ dimes a peck?

4. What will $2\frac{3}{4}$ yards of cloth cost at $2\frac{1}{2}$ dimes a nail?

5. At $2\frac{1}{2}$ dimes a yard, what will be the cost of $13\frac{3}{4}$ yards of muslin?

6. What will $12\frac{2}{7}$ barrels of wine cost at 7 dollars a barrel?

7. A piece of cloth containing $16\frac{1}{2}$ yards is equally divided between 3 persons: how much has each one?

8. If the twelve months were of equal length, how many days would each contain?

9. If a man travel $2\frac{1}{2}$ miles in $\frac{5}{6}$ of an hour, how far will he travel in 5 hours?

10. If wheat is one dollar a bushel, how much will 1 quart cost?

11. If cloth is 8 dollars a yard, what will 1 nail cost?

12. If wine is 4 dollars a gallon, what will 3 pints cost? 1 quart? 2 gills?

13. What will 9 yards of cloth cost at three cents a nail? At 5 cents? At 7 cents?

14. If 1 quarter of a yard of cloth costs $3\frac{1}{5}$ dollars, what will 4 yards cost?

15. If a man spends $\frac{3}{8}$ of a dollar in $\frac{1}{3}$ of a day, how much will he spend in 2 weeks?

16. If $\frac{3}{7}$ of a hogshead of wine cost 54 dollars, what does the wine cost a gallon?

17. If a man earns 13 dollars a week, how much does he earn in each of the 6 working days?

18. If a man earns $1\frac{3}{4}$ dollars a day, how much does he earn in a month of 26 working days?

19. If hay is 15 dollars a ton, how much is that per 1*cwt.*? For 1 quarter?

20. If 10 pounds of hay cost $5\frac{1}{2}$ mills, how much will a ton cost?

LESSON VI.

Per Cent and Per Centage.

1. What is 1 per cent of 1 dollar? What is 2 per cent of 1 dollar? 3 per cent? 4 per cent? 5 per cent?*

2. What is 4 per cent of 50?

3. What is 6 per cent of 200 dollars?

4. What is 4 per cent of 150 dollars? Of 200 dollars?

---

* Suggestion.—Per cent means by the hundred. Thus, 1 per cent, 2 per cent, 3 per cent, 4 per cent, &c., of any number or

5. What is 9 per cent of 300 dollars? Of 400 dollars?

6. What part of 1 is 3 per cent of 30? Of 4 per cent of 25?

7. What is 6 per cent of 60 dollars? Of 50 dollars?

8. What is 8 per cent of 70 dollars? Of 90 dollars?

9. What is 4 per cent of 250 dollars? Of 45 dollars?

10. What is 9 per cent of 40? Of 50? Of 60?

11. What is $3\frac{1}{2}$ per cent of 100 dollars? Of 40 dollars?

12. What is $2\frac{1}{2}$ per cent of 200 dollars? Of 60 dollars?

13. What is 4 per cent of 600 dollars? Of 30 dollars?

14. What is 3 per cent of 25 dollars? Of 36 dollars?

15. A person has 250 dollars, and takes out $2\frac{1}{2}$ per cent: how much will he have left?

16. What is 8 per cent of 15 dollars? Of 25? Of 30? Of 45?

---

thing, means that the number or thing is divided into 100 equal parts, and that 1, 2, 3, 4, &c., of these parts are taken. The number of parts taken, determines the rate per cent. Thus, the rates above, are 1, 2, 3, 4, &c., per cent.

The part of the number taken, is called the *per centage*. Thus, when the thing is 1 dollar, and the rate 1 per cent, 1 cent is the per centage; if the rate is 2 per cent, the per centage is 2 cents, &c. Observe,

If any number be divided by 100, *the quotient will be* 1 *per cent of that number.* Hence, to find the per centage of any number, for any rate per cent.

Multiply the given number by the rate per cent and cut off two figures from the right hand of the product, which is equivalent to dividing by 100.

17. What is 10 per cent of 20 dollars? Of 25? Of 35? Of 40?

18. What is the per centage of 75 dollars, at the rate of 6 per cent?

19. What is the per centage of 80 dollars, at the rate of 5 per cent?

20. What is the 8 per cent of a piece of cloth, measuring 50 yards?

21. What is $3\frac{1}{2}$ per cent of a piece of muslin, measuring 75 yards?

22. What is 20 per centage of a box of shoes, containing 250 pairs?

23. What part of a number is 5 per cent of that number?*

24. Forty per cent is what part of any number?

25. What is 25 per cent of 60? Of 50? Of 40?

---

* ANALYSIS—Five per cent of any number, is 5 hundredths of that number: hence 5 per cent of a number being $\frac{5}{100}$ of that number $= \frac{1}{20}$ of that number. The following table shows the per centage in terms of the number:

5 per cent equals $\frac{5}{100} = \frac{1}{20}$ of the number
6 per cent equals $\frac{6}{100} = \frac{3}{50}$ of the number
10 per cent equals $\frac{10}{100} = \frac{1}{10}$ of the number.
$12\frac{1}{2}$ per cent equals $\frac{25}{200} = \frac{1}{8}$ of the number.
15 per cent equals $\frac{15}{100} = \frac{3}{20}$ of the number.
20 per cent equals $\frac{20}{100} = \frac{1}{5}$ of the number.
25 per cent equals $\frac{25}{100} = \frac{1}{4}$ of the number.
30 per cent equals $\frac{30}{100} = \frac{3}{10}$ of the number.
$33\frac{1}{3}$ per cent equals $\frac{100}{300} = \frac{1}{3}$ of the number.
$37\frac{1}{2}$ per cent equals $\frac{75}{200} = \frac{3}{8}$ of the number.
40 per cent equals $\frac{40}{100} = \frac{2}{5}$ of the number.
50 per cent equals $\frac{50}{100} = \frac{1}{2}$ of the number
75 per cent equals $\frac{75}{100} = \frac{3}{4}$ of the number.
100 per cent equals $\frac{100}{100} = 1$ the number.

26. What is 10 per cent of 60? Of 40? Of 15?

27. What is 5 per cent of 40? Of 80? Of 100?

28. What is 15 per cent of 40 dollars? Of 80 dollars?

29. What per cent of any number is $\frac{1}{5}$ of it? what per cent is $\frac{1}{6}$ of it?*

30. Five is what per cent of 20?†

31. Six is what per cent of 18? Of 24? Of 30?

32. Ten is what per cent of 50? Of 30? Of 40? Of 60?

33. Three is what per cent of 12? Of 15? Of 24? Of 36?

34. Forty is what per cent of 80? Of 20? Of 10?

35. Fifty is what per cent of 200? Of 60? Of 100?

36. Seven is what per cent of 49? Of 21? Of 56?

37. Eight is what per cent of 56? Of 64? Of 84?

---

* ANALYSIS.—Since $\frac{1}{5}$ of a number equals $\frac{20}{100}$, it follows that $\frac{1}{5}$ of a number is equal to 20 per cent; and we may find the 20 per cent by adding two 0's to 1, and then dividing by 5. Hence, we see, that having written the per centage in the form of a fraction, *if we add two cyphers to the numerator and then divide, the quotient will express the rate per cent.*

Therefore, the rate per cent, when the per centage is $\frac{1}{6}$, is 100 divided by 6, which gives $16\frac{2}{3}$ per cent.

† ANALYSIS.—Five is what part of 20? (see lesson VI, page 89.) 5 is $\frac{5}{20}$ of 20 $= \frac{1}{4}$ of 20; but $\frac{1}{4}$ is 25 per cent: hence, 5 is 25 per cent of 20.

Hence, to find the rate per cent when the per centage and number are known: *Divide the per centage by the number.*

QUESTIONS.

1. A grocer purchased a bag of coffee at ten cents a pound: at what price must he sell it a pound, in order to make 10 per cent? What must he sell it for, to make 25 per cent? 50 per cent?

2. If a piece of broadcloth, containing 30 yards cost 5 dollars a yard, what must it be sold for to gain 20 per cent? What will be the profit?

3. A grocer bought 10 barrels of flour, at 8 dollars a barrel: what must they be sold for, to gain 25 per cent?

4. If sugar is bought at 6 cents a pound, what per cent will be gained if it be sold at 7?

5. If a barrel of flour cost 8 dollars, what must it be sold for, to gain 5 per cent?

6. A merchant finds that a lot of goods, which cost 60 dollars, is damaged, and he sells them at a loss of 15 per cent: what does he get for them?

7. The price of a book is 80 cents; but being sold to a friend, a discount is made of 20 per cent: what is paid for it?

8. A piece of cloth, which cost $45, is somewhat damaged, and is sold at a discount of $33\frac{1}{3}$ per cent: what is paid for it?

9. A merchant buys a chest of tea, for which he pays 85 dollars; but finds it injured, and sells it at 20 per cent loss: how much does he get for it?

10. If a grocery merchant buys sugar at 6 cents a pound, and sells for 8, what per cent does he make?

11. A merchant buys a barrel of sugar for $60 and sells it for $80: what was the rate per cent and what the percentage?

12. A grocer buys sugar at 5 cents a pound: what must he sell it for to make 60 per cent?

13. A grocer buys sugar at 8 cents a pound: what must he sell it for to make 25 per cent?

14. A grocer buys sugar at 6 cents a pound and sells it at 9: how much does he make per cent?

15. A grocer buys a bag of coffee at 12 cents a pound: what must he sell it for a pound in order to net $16\frac{2}{3}$ per cent?

16. A bag of coffee is bought at 10 cents a pound, nd being injured, is sold at 8 cents a pound: what was the loss per cent?

17. If flour cost 9 dollars a barrel, what must it be sold for to give 10 per cent. profit? $12\frac{1}{2}$ per cent profit? 18 per cent?

18. If molasses costs 30 cents a gallon, what must it be sold for to yield a profit of 20 per cent? $33\frac{1}{3}$ per cent?

19. What is 25 per cent of 6? Of 9? Of 10?

20. Nine is what per cent of 36? Of 54?

LESSON VII.

Of Interest.

1. Interest is an allowance made for the use of money, and is generally reckoned at so much per cent for each year on the sum loaned, which sum is called the *principal*. The allowance, or per centage, is called the interest, and the principal and interest, together, are called the *amount*.

2. What is the interest of $100 for 1 year, at 1 per cent? At 2 per cent? At 3 per cent? At 4 per cent?

3. What is the interest of $150 for 1 year, at 1 per cent? At 2 per cent? At 3 per cent?

4. What is the interest of $200 for 1 year at 2 per cent? At 5 per cent?

5. What is the interest of $160 for 1 year, at 5 per cent? At 8 per cent?

6. What is the interest of $200 for 2 years at 5 per cent?*

7. What is the interest of $200 for 1 year at 6 per cent?

8. What is the interest of $150 for 1 year at 8 per cent?

9. What is the interest of $300 for 2 years at 4 per cent?

10. What is the interest of $500 for 2 years at 6 per cent?

11. What is the interest of $90 for 3 years at 2 per cent?

12. What will be the amount if $120 be put at interest for 2 years at 6 per cent?

13. What will be the amount if $60 be put at interest for 3 years at 3 per cent?

14. If $80 be put at interest for 2 years at 4 per cent, what will be the amount?

15. What will be the amount of $70 for 2 years at 5 per cent?

16. What is the interest of $320 for 4 years at 3 per cent?

17. What is the interest of $260 for 3 years at 6 per cent?

18. What is the interest of 125 dollars for 2 years at 7 per cent?

19. What will be the amount, if $300 be put at interest for 2 years, at the rate of 4 per cent?

---

* ANALYSIS.—The interest of 1 dollar for 1 year at 1 per cent is 1 cent; and for any number of dollars, as many cents as there are dollars in the principal. Hence:

Find the interest of the principal for 1 year, at one per cent and then multiply by the time, in years, and by the rate per cent: the product will be the interest.

Thus, the interest of $200 for 1 year, at 1 per cent, is $2: then, $2 × 2 × 5 = $20, the interest for 2 years, at 5 per cent.

☞ The tenths of a dollar may be read dimes, and the hundredths, cents.

20. What will be the amount, if $250 be put at interest at 6 per cent for 3 years?

21. What will be the amount, if $500 be put at interest for 2 years at 4 per cent?

22. What will be the amount, if $55 be put at interest for 5 years at 7 per cent?

23. What will be the interest of $85 for 2 years at 7 per cent?

24. What will be the interest of $75 at 6 per cent for 5 years?

25. What will be the interest of $275 at 6 per cent for 4 years?

26. What will be the amount of $175, after drawing interest for 3 years at 5 per cent?

27. What will be the interest of $160 for 2 years at 8 per cent?

28. What will be the interest of $375 for 2 years at the rate of 5 per cent?

29. What will be the amount of $350, drawing interest for 2 years at the rate of 4 per cent?

30. What will be the amount of $95 for 2 years at the rate of 5 per cent?

31. What will be the amount of $86 for 3 years at the rate of 3 per cent?

32. What is the interest of $150 for 3 years at the rate of 6 per cent?

33. What is the amount of $360 for 5 years at 2 per cent?

34. What is the amount of $240 for 3 years at 3 per cent?

35. What will be the amount of $120 for 4 years at the rate of 5 per cent?

36. What will be the amount of $240 for 3 years at 5 per cent?

LESSON VI.

Interest for parts of a Year.

1. What is the interest of 200 dollars for 3 months, at 5 per cent?* For 5 months?

2. What is the interest of 60 dollars for 4 months, at 5 per cent?

3. What is the interest of $40 for 6 months at 6 per cent?

4. What is the interest of $20 for 9 months, at 3 per cent?

5. What is the interest of $15 for 10 months, at 6 per cent?

6. What is the interest of $6 for 8 months, at 7 per cent?

7. What is the interest of $12 for 11 months, at 8 per cent?

8. What is the interest of $25 for 10 months, at 9 per cent?

9. What is the interest of $60 for 5 months, at 6 per cent?

10. What is the interest of $84 for 11 months at 10 per cent?

11. What is the interest of $96 for 7 months, at 9 per cent?

12. What is the interest of $72 for 5 months, at 7 per cent?

13. What is the interest of $144 for 11 months, at 9 per cent?

---

* ANALYSIS.—Find the interest for 1 year, at 1 per cent, which is $2; then, as 3 months being $\frac{1}{4}$ of a year, the interest for three months will be $\frac{1}{4}$ of 2, or $\frac{1}{2}$ a dollar; then multiply by 5, the rate per cent, and we obtain $2\frac{1}{2}$ dollars, or 2 dollars and fifty cents, for the interest.

For 5 months, we have $\frac{5}{12}$ of 2 dollars, which is $\frac{5}{6}$ of one dollar, which being multiplied by 5, gives $\frac{25}{6}$, or $1\frac{1}{6}$ dollars.

14. What is the interest of $60 for 6 days,* at 9 per cent?

15. What is the interest of $84 for 9 days, at 7 per cent?

16. What is the interest of $24 for 16 days, at 8 per cent?†

17. What is the interest of $48 for 17 days, at 6 per cent?

18. What is the interest of $50 for 6 days, at 8 per cent?

19. What is the interest of $96 for 11 days, at 7 per cent?

20. What is the interest of $40 for 15 days, at 9 per cent?

21. What is the interest of $144 for 8 days, at 5 per cent?

22. What is the interest of $60 for 18 days, at 10 per cent?

23. What is the interest of $72 for 10 days, at 9 per cent?

24. What is the interest of $132 for 5 days, at 6 per cent?

25. What is the interest of $42 for 20 days, at 9 per cent?

26. What is the interest of $12 for 19 days, at 10 per cent?

27. What is the interest of $36 for 21 days, at 10 per cent?

28. What is the interest of $15 for 25 days, at 8 per cent?

---

* ANALYSIS.—The interest of $60 for 1 year, at 1 per cent is 60 cents; and for 1 month, is 5 cents, and for 1 day is $\frac{5}{30}$ or $\frac{1}{6}$ of 1 cent; and at 8 per cent., it is $\frac{1}{6} \times 8 = \frac{8}{6} = 1\frac{1}{3}$ cents

† OBSERVATION.—Observe that 16 days is $\frac{1}{2}$ of a month, and 1 day over; then as the interest at 1 per cent is 2 cents a month, 16 days gives 1 cent, and $\frac{1}{30}$ of 2 cents, or $\frac{1}{15}$ of a cent: hence $1\frac{1}{15}$ of a cent multiplied by 8, gives $8\frac{8}{15}$ of a cent.

29. What is the interest of $20 for 2 years 3 months and 6 days, at 7 per cent?

30. What is the interest of $30 for 3 years 9 months and 10 days, at 6 per cent?

31. What is the interest of $24 for 4 years 6 months and 20 days, at 8 per cent?

32. What is the interest of $36 for 1 year and 5 days, at 8 per cent?

33. What is the interest of 60 dollars for 2 years 8 months and 25 days, at 6 per cent?

34. What is the interest of $200 for 2 years 10 months and 10 days, at 5 per cent?

35. What is the interest of 60 dollars for 2 years 2 months and 3 days, at 8 per cent?

36. What is the interest of 48 dollars for 3 years 8 months and 10 days, at 2 per cent?

37. What is the interest of $72 for 3 years 6 months and 5 days, at 4 per cent?

38. What is the interest of $84 for 2 years 4 months and 3 days, at 6 per cent?

LESSON VIII.

Analysis of Questions.

1. If 1 yard of cloth cost $2, how much will 4 yards cost at the same rate?

2. If 5 yards of cloth cost $10, what will 8 yards cost at the same rate?

3. If 4 yards of cloth cost $9, what will 16 yards cost at the same rate?*

---

* ANALYSIS.—This, and similar examples, may be done by the analysis on page 53; or they may be done thus:

The ratio of 4 yards of cloth to 16 yards is 4; that is, 16 yards of cloth is 4 times as much as 4 yards, and therefore, will cost 4 times as much?

If 4 yards of cloth cost $9, 16 yards will cost 4 times 9 dollars, which are 36 dollars.

4. If 6 men consume a barrel of flour in 2 months, how much will they consume in a year?

5. If a man travels 60 miles in 5 days, how far will he travel in 30 days?

6. If 4 men consume 1 barrel of flour in 20 days, how much would 32 men consume in the same time?

7. If 3 barrels of flour cost $14, how much will 9 barrels cost?

8. If 4*lb.* of sugar cost 64 cents, what will 13*lb.* cost?

9. If $\frac{3}{4}$ of a piece of cloth costs $$8\frac{1}{4}$, what will $\frac{9}{4}$ pieces cost?

10. If $\frac{4}{7}$ of a barrel of cider cost $\frac{9}{11}$ of a dollar, what will $\frac{2}{7}$ of a barrel cost?

11. If 9 bushels of oats will feed 4 horses 5 days, how long will 36 bushels feed them?

12. If 3 paces of the common step be equal to 2 yards, to how many yards will 18 paces be equal?

13. If 5 yards of cotton cloth are equal in value to 2 yards of linen, how many yards of linen will 20 yards of cotton buy?

14. If 8*lb.* of coffee is of the same value as 3*lb.* of tea, how many pounds of tea will 24*lb.* of coffee buy?

15. If 8 oranges are worth 24 cents, how much are 2 oranges worth?

16. If 4 apples are worth $1\frac{1}{2}$ oranges, and one orange is worth two lemons, how many lemons will 12 apples buy?

17. If 5 baskets of peaches are worth $$8\frac{3}{4}$, how much will 8 baskets be worth?

18. If a man travel 7 miles in two hours, how far will he travel in 14 hours?

19. Two men start from the same point and travel in opposite directions; one at the rate of $3\frac{1}{2}$ miles an hour, and the other at the rate of $4\frac{1}{4}$ miles an

hour; how far apart will they be at the end of 4 hours?

20. If, in the last question, the men were to travel in the same direction, how far apart would they be at the expiration of 4 hours?

21. If 6*lb.* of butter cost 14 shillings, how much will 15*lb.* cost?

22. If 4 turkeys cost $\$2\frac{3}{4}$, what will 16 turkeys cost?

23. If 9 yards of broadcloth cost $\$27\frac{1}{3}$, how much will 27 yards cost?

24. If 5*lb.* of loaf sugar cost $\$\frac{3}{5}$, how much will twenty-five pounds cost?

25. If 2 tons of hay cost $19\frac{1}{2}$ dollars, what will 8 tons cost?

26. If 3 pairs of boots cost $\$21\frac{1}{3}$ dollars, how much will 18 pairs cost?

27. If 3 horses eat $6\frac{1}{4}$ bushels of oats in 2 days, how much will 12 horses eat in the same time?

28. If a family consume $1\frac{3}{4}$ barrels of flour in 2 months, how much will they consume in 4 months?

29. If 7 bushels of wheat cost $\$6\frac{3}{4}$, how much will 14 bushels cost?

30. If a family consume $2\frac{2}{5}$ bushels of grain in $3\frac{1}{4}$ weeks, how much will they consume in $6\frac{1}{2}$ weeks?

31. If $3\frac{1}{2}$ yards of cloth cost $\$6\frac{2}{5}$, what will 14 yards cost?

32. If 3 pairs of shoes cost $\$5\frac{3}{4}$, what will 9 pairs cost?

33. If a man travels 9 miles in $2\frac{1}{2}$ hours, how far will he travel in $7\frac{1}{2}$ hours?

34. If $9\frac{1}{3}$ pounds of tea cost $12\frac{4}{5}$ dollars, how much will 28 pounds cost?

35. If 8 yards of broadcloth cost 17 dollars, what will 16 yards cost?

LESSON IX

Analysis of Questions Continued.

1. If a man can build a wall in 1 day, how long will it take two men to build it?

2. If 2 men can build a wall in 4 days, how long will it take 4 men to build it?

3. If 4 horses, in two days, eat 5 bushels of oats, how much will 6 horses eat in 4 days?*

4. If a barrel of flour last 15 men 20 days, how long will it last 25 men?†

5. If 6 men consume 24*lb.* of beef in 5 days, how much will 9 men consume in ten days?

6. If 6 horses eat $2\frac{1}{2}$ tons of hay in 2 weeks, how much will 16 horses eat in $1\frac{1}{2}$ weeks?

7. If 8 men can build a wall 6 days, in how many days can 12 men build it?

8. If a certain amount of provisions will last 2 families of 5 persons each 3 weeks, how long will the same provisions last 5 families of 6 persons each?

9. If 2 men, in 5 days, can build 160 feet of wall, how long will it take 4 men to build 192 feet of wall?

10. If 5 men can do a certain work in 6 days, how long will it take 3 men to do 5 times that work?

---

* Analysis—Four horses will eat as much in 2 days as 8 horses in 1 day; and 6 horses will eat as much in 4 days as 24 horses in 1 day. Now, 24 horses is 3 times 8 horses: if 8 horses eat 5 bushels of oats in 1 day, 24 horses will eat 3 times as much, which are 15 bushels: therefore, 6 horses in 4 days (equivalent to 24 horses for 1 day) will eat 15 bushels.

† Fifteen men will eat as much in 1 day as 1 man will eat in 15 days and 15 men will eat as much in 20 days as 1 man in 300 days. Now, the same provisions will last 25 men only one-twenty-fifth as long as they will last 1 man; that is, so many days as 25 is contained times in 300, which are 12.

11. If a man travels 48 miles in 2 days, travelling 6 hours a day, how far will he travel in 3 days, travelling at the same rate, 5 hours a day?

12. If 12 dollars' worth of provisions will supply 9 men 4 days, how much will it cost to supply 21 men for 5 days?

13. If 2 men consume *2lb. 4oz.* of flour in 1 day, how much will 8 men consume in 4 days?

14. How many sheep, at 3 dollars a head, must be given for 5 cows at $18 apiece?

15. A man, failing in trade, pays his creditors 3 shillings on the dollar, while another pays an equal sum by paying two shillings on the dollar: what is the ratio of their debts?

16. If $\frac{5}{6}$ of a bushel of oats will feed 2 horses half a day, how many will be required to feed 4 horses $4\frac{1}{2}$ days?

17. If a barrel of flour will serve a family of 6 persons $3\frac{1}{2}$ weeks, how much will serve a family of 9 persons $4\frac{2}{3}$ weeks?

18. If 5 men can cut 30 cords of wood in 3 days, how much will 4 men cut in 8 days?

19. If 3 men can mow 15 acres in $2\frac{1}{2}$ days, how long will it take 11 men to mow 44 acres?

20. If a farmer can plough 9 acres in 4 days, with one team, how much can he plough with two teams in 8 days?

21. If a pasture of 8 acres will feed 3 horses for 2 months, how many acres will feed 4 horses 5 months?

22. If the wages of 3 men for 7 days are 21 dollars, what will be the wages of 9 men for 11 days?

23. If a man travels $12\frac{1}{2}$ miles in 5 hours, how far will he travel in 4 hours, at the same rate?

24. A man loses, at play, $\frac{3}{7}$ of his money, after which he gives away $\frac{1}{4}$ of the remainder, and finds that he has 8 dollars left; how much had he at first?

25. How many yards of cloth at $4\frac{1}{2}$ dollars a yard, must be given for 9 yards at 3 dollars a yard?

26. If 4 tailors can make 8 pair of pantaloons in 2 days, how many will 3 tailors make in 5 days?

27. If 5 horses consume $24\frac{1}{2}$ tons of hay in a winter, how much will 10 horses consume?

28. James being asked how many marbles he had, replied, that if $\frac{3}{5}$ of the number be divided by 2, and the quotient subtracted from one half the number, the remainder would be 12: how many had he?

29. If 9 men can do a piece of work in $4\frac{2}{3}$ days, how many men should be employed to do the same work in 7 days?

30. If 3 horses eat $3\frac{1}{2}$ tons of hay in 2 months, how much will supply 5 horses for 4 months?

LESSON X.

To find the parts, knowing the whole and the proportion of the parts.

1. James bought an orange and a melon, for which he paid 8 cents. He paid three times as much for the melon as for the orange: what did he pay for each?*

2. Charles bought a whistle and a top, for which he paid 12 cents. He paid five times as much for the whistle as for the top: what did he pay for each?

3. What number added to itself will give a sum equal to 20?

4. What number added to twice itself will give a number equal to 15?

---

* ANALYSIS.—James paid *one equal part of the whole sum* for the orange, and 3 *equal parts* for the melon: hence, he paid 4 *equal parts* for both. Then, the whole sum which he paid, (8 cents), divided by the number of equal parts, (4 equal parts), will give 1 part, which is 2 cents, what he paid for the orange; and 2 cents multiplied by 3, will give 6 cents, what he paid for the melon.

5

5. What number added to five times itself will give a number equal to 30?

6. John bought an apple, a peach, and an orange, for which he paid 6 cents. He paid twice as much for the peach as for the apple, and as much for the orange as for the apple and peach together: what did e pay for each?

7. A man bought a horse, saddle, and bridle, for which he paid 90 dollars. He paid twice as much for the saddle as for the bridle, and four times as much for the horse as for the saddle and bridle together: what did he pay for each? How many parts are there?

8. The sum of the ages of James, Charles, and John, is 44 years. James' age is one half Charles and one third John's: what is the age of each?

9. A farmer has in his garden apple trees, pears, and peaches; in all 72. He has twice as many pear as apple trees, and three times as many peaches as pear trees: how many has he of each?

10. A person distributed 36 cents amongst three beggars, a father, mother, and son. He gave the mother twice as much as the boy, and the father twice as much as the mother and boy together: how much did he give to each?

11. A farmer has 72 sheep in four lots. In the second he has twice as many as in the first; in the third as many as in the second; and in the fourth twice as many as in the third: how many has he in each?

12. Mary, Jane, and Nancy, gather 144 apples from the orchard. Jane is to have twice as many as Mary, and Nancy is to have three times as many as Mary and Jane together: how many will each have?

13. Divide 12 into two such parts that the second shall be double the first. Into how many *equal* parts is 12 to be divided?

14. Divide 21 into three such parts that the second shall be double the first, and the third double the second.

15. James asked John how many marbles he had? John replied, if you will give me twice as many as I now have, and William will give me 5 times as many as I should then have, I would have, in all, 36: how many had he?

16. Mr. Parsons bought 4 pounds of coffee, a pound of tea, and a yard of cloth, and paid in all $16. He paid twice as much for the tea as for the coffee, and 5 times as much for the cloth as for the coffee: what did he pay for each?

17. Divide 36 into four such parts that the second shall be 3 times the first, the third 5 times the first, and the fourth 9 times the first.

18. In a pasture there are seven times as many sheep as cows, and twice as many lambs as cows; in all, 40: how many of each sort?

19. A father, mother, and son, receive 108 cents for a day's work. The mother receives twice as much as the son, and the father as much as the mother and son together: what does each receive?

20. In a school of three departments, there are 150 pupils. In the first department there are one third as many as in the second, and in the second one half as many as in the third: how many are there in each? How many equal parts of the whole school in each department?

21. Divide 82 into four such parts that the second shall be 4 times the first, the third 3 times the second, and the fourth 2 times the third.

22. A sloop employed in carrying bricks to New York, carries 40 thousand at a load. She is loaded twice and unloaded once, the first week—the second week, she is unloaded twice and loaded once; and so

on for the season: how much does she average a week?

23. James being asked what he had been about during the day, replied, that he had been ciphering 4 hours and done 82 sums. That in the second hour he did 4 times as many as in the first; in the third hour, three times as many as in the second; and in the fourth, 2 times as many as in the third: how many did he do in each hour?

LESSON XI.

To divide Numbers into proportional parts.

1. Divide the number 18 into two parts, such that the ratio of the parts shall be the same as 2 to 4.*

2. Divide the number 28 into two parts, such that their ratio shall be the same as 5 to 9.

3. Divide the number 34 into two parts, such that the first shall be eight ninths of the second.

4. Divide 34 into two such parts that the first shall be one and one eighth times the second.

5. Two men bought a piece of muslin containing 30 yards; one paid \$2 and the other \$3: how many yards belonged to each?

6. Two men hired a pasture for \$24. One pastured 5 horses and the other 3: how much should each pay?

7. Two men hired a pasture for \$72. One pastured 3 horses for 5 weeks, and the other 7 horses for 3 weeks: what proportion should each pay?

8. A father divides 84 cents between John and Charles, giving 5 cents to John and 7 to Charles each

* ANALYSIS.—There are 6 units in the sum of 2 and 4. If 18 be divided into 6 equal parts, each part will be 3. Two of these parts must form the first number, and 4 of them the second. Hence, the numbers are 6 and 12.

time, till the whole was distributed: how much did he give to each?

9. Three persons buy a piece of cloth containing 48 yards. The first puts in 5 dollars, the second 9, and the third 10: what was each one's share?

10. James has 72 marbles; he gives 3 to William and 5 to John, each time, until none are left: how many does he give to each?

11. William has 9 cents and John 7, and they buy 96 apples: how many apples should each have?

12. Mr. Wilson fails in business and pays $\frac{7}{9}$ of his debts: how much will Mr. Squires receive to whom he owes $108?

13. A grocer weighs out 24 pounds of sugar to 2 customers, giving 2 pounds to one as often as he gave $\frac{2}{3}$ of a pound to the other: how much did he give to each?

14. A draper divides a piece of cloth containing 36 yards, between 2 persons, giving $2\frac{1}{2}$ yards to the one every time that he gave $3\frac{1}{2}$ yards to the other: how much did each receive?

15. A man distributed 78 cents among 8 beggars, 8 of whom were men and 5 were women. He gave twice as much to each woman as to each man: how much did he give to each?

16. James and John start from the same place, travel the same way, and take steps of equal length James steps 4 times while John steps but 3: how far will they be apart when the distances travelled by both is 14 miles?

17. James and John start from the same place, travel the same way, and take steps of equal length; James steps 4 times while John steps but 3: how far will each have travelled when they are 3 miles apart?

18. Two men agree to do a piece of work for which they are to receive $88; the first sends 4 hands for 3 days, and the second 5 hands for 2 days: how much should each receive?

19. A person met three beggars, a boy, a mother and father, and distributed 84 cents among them. For every 5 cents he gave the boy he gave the mother 7 and the father 9: how much did he give to each?

20. Three persons hire a pasture, for which they pay $56. The first puts in 2 horses for 3 weeks, the second 5 horses for 2 weeks, and the third 9 horses for $1\frac{1}{3}$ weeks: how much ought each to pay?

21. Charles has 5 marbles and John 9, and they agree to share their winnings or losses in the same proportion. After several games they find that they have won 42: how are they to be divided?

22. A and B enter into partnership: A puts in 7 dollars, and B 11: they make 9 dollars by the operation: how should it be divided?

23. A and B enter into partnership: A puts in 6 dollars for 2 months, and B, 5 dollars for 3 months: they gain 81 dollars: what is the share of each?

24. Two barrels of flour, costing 12 dollars, are consumed by three persons; the first ate from them 2 months, the second 3 months, and the third 5 months: how much should be paid by each?

25. Three persons hire a pasture for sheep, for which they pay 12 dollars. The second puts in twice as many sheep at the first, and the third three times as many as the first; but the sheep belonging to the first man are in twice as long as those belonging to the second, and three times as long as those belonging to the third: how much should each pay?

LESSON XII.

Analysis of Questions.

1. The sum of two numbers is 10 and their difference 4: what are the numbers?*

2. The sum of two numbers is 16 and their difference 8: what are the numbers?

3. James and John together have 24 marbles, and the difference between their marbles is one fourth of the sum: how many had each?

4. A tailor in measuring two pieces of cloth found their difference to be 6 yards, and also that this difference was an eighth part of the cloth in both pieces: how much was there in each piece?

5. James and John have 16 marbles, and James has 4 more than John: how many has each?

6. Nancy has 6 more pins in her cushion than Jane, and together they have 30: how many has each?

7. John, in a week recited 10 lessons more than Charles, and together they recited 24: how many did each recite?

8. A farmer bought an equal quantity of sugar and coffee, and then gave a cheese for 20 pounds of coffee, when it appeared that he had in all 50 pounds of sugar and coffee: how much had he of each?

9. A man being asked how much money he had, said, that he had only dollars and dimes, and that he had 72 pieces in all: that the number of dollars less the number of dimes, was one-twelfth the sum of the pieces: how much money had he?

---

* ANALYSIS.—If the two numbers were equal and their sum 10, each number would be 5. Now, if you take 1 from one of the 5's you have the number 4, and if you add it to the other you get the number 6; if you do the same for the two last numbers, you get the numbers 3 and 7, whose difference is 4: that is, *the greater of two numbers is equal to half their sum plus half their difference, and the less is equal to half their sum minus half their difference.*

10. A farmer has 10 more sheep than he has cows, he loses 3 cows and 6 sheep, when he finds that he has 17 of both kinds remaining: how many had he at first of each kind?

11. A father being asked his age, replied, that 7 years ago his age was double that of his son's, and now that the sum of their ages was 89: what was the age of each?

12. A man and son engage to work 20 days, the son to receive 3 dimes a day less than the father: at the end of the time they receive 42 dollars: how much of this sum did each earn, and what did each receive per day?

13. John being asked how many marbles he had, replied, that 19 was 3 more than $\frac{4}{5}$ of the number: how many had he?

14. Lucy being asked her age, said, that her sister Jane was 6 years old when she was born, and that now the sum of their ages was 20: what was the age of each?

15. A farmer had as many sheep as hogs, and after losing 12 of his hogs, his sheep and hogs amounted to 60: how many had he at first of each kind?

16. A man bought a vest, for which he paid 21 dollars less than he paid for his coat, and for the two together he paid 33 dollars: what did he pay for each?

17. Mr. Wilson sold his cow for 30 dollars, which was $\frac{10}{11}$ of what she cost him: what did he give for her?

18. A gentleman bought a coat and hat, for which he paid 27 dollars, and the cost of the hat was one eighth of the cost of the coat: what was the cost of each?

19. If $\frac{5}{8}$ of a number is 3 less than $\frac{6}{8}$, what is the number?

20. Two men have 60 dollars between them; if the one having the largest sum gives the other 5 dollars, their money will be equal: how much had each?

21. A man bought a cow and a calf, for which he paid 36 dollars, paying 5 times as much for the cow as for the calf: what did he pay for each?

22. A man bought a pair of boots for $6\frac{3}{4}$ dollars, which was $\frac{3}{7}$ of what he paid for his coat: what did his coat cost?

23. A man paid $2\frac{1}{2}$ dollars more for his pantaloons than for his vest, and for both he paid $14\frac{1}{2}$ dollars: what did he pay for each?

24. A market woman bought 36 eggs; for $\frac{3}{4}$ of them she paid 2 cents for three eggs, and the remainder she bought at the rate of 4 cents for 3 eggs: for what must she sell them that she may make 6 cents?

LESSON XIII.

Separate and Concurring Causes.

1. If Charles can do a piece of work in 2 days, what part of it can he do in 1 day?

If you denote the work by 1, what will denote the part which he can do in 1 day?

2. If James can do a piece of work in 3 days, what part of it can he do in 1 day? What part in 2 days?

3. If a family consume 12 pounds of sugar in a week, how much will they consume in 1 day? How much in 3 days? 4 days?

4. If James can do one fifth of a piece of work in 1 day, how long will it take him to do the entire work? If he can do $\frac{1}{4}$ in a day, how long will it take him to do the work? How long if he can do $\frac{1}{3}$? If he can do $\frac{1}{6}$, how long?

5. John can do a piece of work in 2 days, and Charles can do the same work in 3 days:

What part of the work can each do in a single day?

What part can both do in 1 day; and in what time can the work be done by both of them working together?*

6. A can do a certain piece of work in 3 days, B can do it in 5 days:

What part of it can each do in 1 day? What part can both do in 1 day, and in what time can the work be done by both working together?

7. A cistern is to be filled by two pipes. One can fill it in 2 hours and the other in 5:

What part of the cistern can each fill in 1 hour? What part of it can they both fill in 1 hour? In what time can they fill the cistern running together?

8. A cistern is to be filled by three pipes. The first can fill it in 2 hours, the second in 3, and the third in 4:

What part of it will each fill in 1 hour? What part of it will they all fill in 1 hour? In what time will the three fill it, running together?

9. John can do a piece of work in 2 days, and John and James together, can do it in $1\frac{1}{5}$ days: in what time can James do it alone?†

---

* ANALYSIS.—Denote the work to be done by 1. Then $\frac{1}{2}$ will will represent the part which John can do in 1 day; and $\frac{1}{3}$ will represent the part which Charles can do in 1 day; and the sum of $\frac{1}{2}$ and $\frac{1}{3}$ will represent what both can do in 1 day, viz.: $\frac{5}{6}$ of the entire work. Then, the number of times which 1 (the work to be done) contains $\frac{5}{6}$, viz.: $1\frac{1}{5}$ times, shows the number of days in which they can do the work together. Many similar examples may be done in the same manner.

† ANALYSIS.—If John can do the work in 2 days, he can do $\frac{1}{2}$ of it in 1 day; and if James and John together, can do the work in $1\frac{1}{5}$ days, that is $\frac{6}{5}$ days, they can do 1 divided by $\frac{6}{5}$, that, is $\frac{5}{6}$ of it in 1 day: then, James can do $\frac{5}{6}$ of the work less $\frac{1}{2}$ of the work, that is, $\frac{1}{3}$ of the work, in 1 day; therefore, he can do the whole work in three days.

10. A can do a piece of work in 3 days, A and B together, can do it in $1\frac{7}{8}$ days: in what time can B do it working alone?

11. A cistern can be filled by one pipe in 2 hours, and by 2 pipes in $1\frac{3}{7}$ hours: how long will the second pipe require to fill it if running alone?

12. A cistern can be filled by 3 pipes in $1\frac{1}{12}$ hours, one of the pipes can fill it in 2 hours, and another in 3: how long will it require the third to fill it, if running alone?

13. A man and his wife usually drank a gallon of beer in 12 days; but when the man was from home it lasted his wife 30 days: how many days would the man require to drink it?

14. A quantity of flour will last one family six weeks, and the same flour will last another family 3 weeks. It is found that one-third of the flour is spoiled: how long will the remainder last both families?

15. If two families consume a quantity of provisions in $1\frac{1}{3}$ weeks, and one family alone would consume the same provision in 4 weeks: how long would it last the other?

16. A can mow a field in 1 day, B in 2 days, and C in 3 days: in what time can they all mow it, working together?

17. A can mow a field in 2 days, B can mow it in 3 days, but by the aid of C, they can mow it in $\frac{6}{11}$ of 1 day: how long will it take C to mow it alone?

In what time can A and B mow it?

In what time can A and C mow it?

18. Three carpenters can finish a house in 2 months; two of them can do it in $2\frac{1}{2}$ months: how long will it take the third to do it alone?

19. A, with the assistance of B, can build a wall 2 feet wide 3 feet high and 30 feet long in 4 days, but with the assistance of C they can do it in $2\frac{1}{2}$ days:

What part of it can A and B build in 1 day?
What part of it can they all build in 1 day?
What part of it can C alone, build in 1 day?
In how many days would C build it alone?

LESSON XIV.

Analysis of Questions.

1. A laborer engaged to work for 16 days on these conditions: For every day he labored, he was to receive 4 shillings, and for each day that he was idle he was to pay 2 shillings for his board; at the end of the time he received 52 shillings: how many days did he work, and how many days was he idle?*

2. A carpenter took an apprentice on these terms: he paid his father, at the end of each month of 26 working days, 3 shillings a day, for every day the boy worked—charged him 20 shillings for clothes, also, 1 shilling for board, for every idle day; at the end of the time there was 30 shillings due him: how many days was he idle?

3. A merchant bought 50 yards of calico, some of which was damaged, on these terms: He was to pay 3 dimes a yard for all that was perfect, and 1 dime a yard for all that was injured; at the settlement he paid 12 dollars: how many yards were injured?

4. A grocer purchased 30 fowls, a part turkeys and a part of them chickens; for the turkeys he was to pay 11 dimes a piece, and for the chickens 4; he paid in all, 24 dollars and 60 cents: how many were there of each kind?

---

* ANALYSIS.—Had he labored the 16 days, he would have received the 64 shillings. But he received only 52 shillings: hence, he lost 12 shillings by idleness. But as he paid 2 shillings a day for his board, and lost 4 shillings a day in wages, he lost, in all, 6 shillings a day: therefore, the *number* of days he was idle will be expressed by 12 divided by 6, giving 2 idle days: therefore, he worked 14 days.

5. A farmer hired a father and son to do 20 days work between them: the father was to have a dollar for every day he worked, and the son 75 cents; at the end of the time, the amout paid was 17 dollars: how many days did each work?

6. A yard stick is broken into 2 parts, the shorter of which is $\frac{4}{5}$ the length of the longer: what is the length of each piece?*

7. A piece of cloth of 40 yards in length is cut into 2 pieces, such, that the smaller piece is $\frac{3}{7}$ of the larger: What is the length of each piece?

8. What number is that to which if $\frac{3}{5}$ of itself be added, the sum will be 32?

9. A coat and vest cost 24 dollars, and the vest cost $\frac{1}{7}$ as much as the coat: what was the cost of each?

10. A cow and calf are worth 56 dollars, and the calf is worth $\frac{3}{11}$ of the cow: what is the value of each?

11. A pole 16 feet long stands in the mud, water, and air. The part in the mud is $\frac{1}{3}$ of the part in the water, and the part in the air is equal to the other two: what is the length of each part?

12. There is a fish weighing 72 pounds. His head weighs twice as much as his tail, and his body weighs as much as his head and tail together: what is the weight of each part?

13. Divide the number 52 into two such parts, that $\frac{5}{8}$ of the larger part shall be equal to the less part.

---

* ANALYSIS.—The fractional unit, in this question, is one fifth the length of the longer piece. There are 5 of these units in the longer piece and 4 in the shorter: hence, there are 9 in the whole stick, which is 36 inches long. The value of the fractional unit, is, therefore, 36 divided by 9, or 4 inches. Hence, one of the pieces is 20 inches, and the other 16. *All similar questions are solved by finding the fractional unit.*

14. A tailor has 48 yards of cloth in 2 pieces; $\frac{2}{5}$ of the longer piece is equal to $\frac{2}{3}$ of the shorter: how many yards in each piece?*

15. A farmer bought pigs and sheep, in all 33: $\frac{2}{7}$ of the pigs was equal to $\frac{1}{2}$ of the sheep: how many were there of each kind?

16. Three fourths of a son's age is equal to on fourth the age of the father, and the sum of their ages is 80 years: what is the age of each?

17. There are 125 sheep in two fields, $\frac{4}{5}$ of the number in one field being equal $1\frac{1}{5}$ times the number in the other: how many in each field?

18. A man after counting his gains at play, found that he had increased his money by $\frac{1}{2}$ of $\frac{1}{3}$, and that he then had 42 dollars: how much had he at first?

19. The difference of two numbers is 6, and the less number is $\frac{2}{3}$ of the greater: what are the numbers?†

20. A father's age is such, that $\frac{5}{8}$ of it is equal to $1\frac{3}{4}$ the age of his son, and the difference of their ages is 36 years: what is the age of each?

21. What number is that which being diminished by the difference between $\frac{3}{4}$ and $\frac{3}{5}$ of itself, leaves a remainder equal to 34?

22. A flag staff 52 feet long, is so broken by the wind, that $\frac{1}{3}$ of the top piece is equal to $\frac{3}{4}$ of the piece left standing: how long are the pieces?

---

* SUGGESTION.—If $\frac{2}{3}$ of the shorter piece equals $\frac{2}{5}$ of the longer, then $\frac{1}{3}$ of the shorter equals $\frac{2}{10}$ of the longer, and the whole of the shorter piece equals $\frac{6}{10}$ of the longer. Always *find the unit of the smaller in terms of the larger number.*

† ANALYSIS.—Suppose the greater number be divided into 3 equal parts; two such parts, are found in the lesser number: hence, the difference between the numbers is equal to $\frac{1}{3}$ of the greater, which is 6: therefore, the greater number is 18 and the less 12.

23. James was asked how many marbles he had, and replied, I have 55, black and red, and $\frac{3}{7}$ of the black make just as many as $\frac{4}{9}$ of the red. Pray how many have I of each sort?

24. A laborer engaged to work 20 days, and was to receive 9 shillings for every day he worked, and pay 3 shillings a day for his board, every day that he was idle; in settling, he received 84 shillings: how many days did he work?

25. A church, including the steeple, is 188 feet high. If the height of the steeple is equal to $\frac{1}{3}$ the height of the body of the building, what is the height of each?

LESSON XV.

Analysis of Questions.

1. William chases Henry, who is 42 feet in advance, around a circular walk of 100 feet. Their steps are each 3 feet, but William takes 6 steps while Henry takes but 5: how many steps must William make to overtake Henry?

2. A hare is 25 of his own leaps before a greyhound, which is pursuing him. The greyhound makes 2 leaps while the hare makes 5; but 1 leap of the greyhound is equal to 3 of the hare's: how many leaps will the greyhound make before he overtakes the hare?*

3. James is in pursuit of John, and 12 of John's steps behind him. James steps 3 times while John

---

* ANALYSIS.—Since the greyhound makes 2 leaps while the hare makes 5, in the time that the greyhound makes 1 leap the hare will make $2\frac{1}{2}$ leaps. But 1 leap of the greyhound is equal to 3 leaps of the hare; hence, every time the greyhound jumps he will gain on the hare $\frac{1}{2}$ of a hare's leap: therefore, he must make 50 leaps to overtake him.

steps 4 times; but James' steps are twice as long as John's: how many steps must James make to overtake him?

4. Mary is 24 years old and Jane is 4: how many years must elapse before Mary's age will be just double Jane's?*

5. A father is 60 years old and his son is 35: how long since the age of the father was double that of the son? What was then the age of each?

6. A mother is 36 years old and her daughter 12: how long before the age of the mother will be double that of the daughter? What will then be the age of each?

7. A mother is 48 years old and her daughter 30: how long since the age of the daughter was half that of the mother? What was then the age of each?

8. A mother is 48 years old and her daughter 10: how long before the age of the daughter will be one-third that of the mother?† What will then be the age of each?

9. A mother is 54 years old and her daughter 20: how long since the age of the daughter was one-third that of the mother? What was then the age of each?

---

* ANALYSIS.—At Jane's birth, 4 years ago, Mary was 20 years old, and Jane's age was 0. Twenty years from that date, Mary will be twice as old as Jane: Hence, *if Mary's age be diminished by Jane's, and the remainder be multiplied by 2, the product will denote Mary's age when it is double Jane's.*

† ANALYSIS.—At the birth of the daughter, the mother was 38 years old. What number added to 38 years will give a sum equal to 3 times the number added?

If 3 times the number added is equal to 38 plus the number, then, twice the number added will be equal to 38; and the number must be one half of 38, or 19 years. Hence, *if the daughter's age be subtracted from the mother's, and the remainder divided by 2, the quotient will be the daughter's age, when it is one-third the mother's.* If the difference be divided by 3, the quotient will be the daughter's age when it is one-fouth the mother's; and so on.

10. A father's age is 45 years, and his son 9 : how long before the age of the son will be one fourth that of his father? What will then be the age of each?

11. A father is 54 years old, and his son 30 : how long since the age of the father was 4 times that of the son? What was then the age of each?

LESSON XVI.

Analysis of Questions by means of Unity.

We have said, that any number, regarded as a whole, may be called Unity.

What may any number be called, when it is regarded as a whole?

1. What number added to twice itself will give a sum equal to 12?*

2. What number is that which added to three times itself will give a sum equal to 24?

3. What number is that which added to half itself will give 6?

4. What number added to half itself will give 9?

5. What number added to one-fourth of itself will give 20?

6. What number is that which added to twice itself, and the sum to 3 times itself will give 30?

7. What number added to half of itself, and to one fourth of itself will give 28?

8. James being asked his age, said, I am half the age of my father, and the sum of our ages is 60 years: What is the age of each?

---

* Analysis.—Call the number sought, unity. Then by the conditions of the question, unity plus twice unity is equal to 12. But unity plus twice unity is equal to 3 times unity, which is equal to 12. Then, if 3 times unity is equal to 12, once unity is equal to 12 divided by 3; which is 4.

Suggestion.—Let the pupil see if 4 will fulfil the conditions: $4 + 4 \times 2 = 4 + 8 = 12$. Let every question of the lesson be analyzed and verified in a similar way

9. What number is that to which if its $\frac{1}{6}$ be added, the sum will be 35 ?

10. A man being asked his age, replied, if to my age you add one-third of it and then $\frac{1}{4}$ of it, the sum will be 57 : How old was he ?

11. John being asked how many marbles he had, said, that one half of what he had, increased by $\frac{1}{3}$, and diminished by $\frac{1}{4}$, was equal to 14.

12. Divide the number 12 in two such parts, that one shall be three times the other ?

13. Divide 24 into 3 such parts, that the second shall be 3 times the first, and the third 4 times the first: What are the numbers ?

14. Divide 32 into 3 such parts, that the second shall be 4 times the first, and the third 3 times the first: What are the numbers ?

15. Divide 27 into two such parts, that the second shall be $\frac{4}{5}$ of the first ?

16. Divide 20 into three such parts, that the second shall be one half of the first, and the third, one third of the second ?

17. James asked Robert how many marbles he had. Robert answered, "If you will give me half as many as I now have, and afterwards give me $\frac{1}{8}$ of what I shall then have, my number will be 54." How many had he ?

18. A young chap asked an old gentleman his age, who replied, "When you was born, I was $\frac{6}{7}$ of my present age : one third of your age plus one fourth of it, is equal to 7 years. Can you now tell how old I am ?"

19. A tailor buys a piece of cloth for 6 dollars a yard, and a piece of equal length for \$2 a yard, and sells the whole at \$4 a yard : does he make or lose ?

20. John and James together have 45 marbles : if John's is equal to $\frac{2}{3}$ of James, how many has each ?

21. A market woman bought geese and turkeys; 36 in all; one seventh of one sort was equal to one half of the other: how many of each kind?

22. A pole 72 feet long has one half as much in the mud as in the water, and twice as much in the air as in the mud and water together: how many feet in each?

23. What number added to $\frac{1}{2}$, to $\frac{1}{3}$, and to $\frac{1}{4}$ of itself will give a sum equal to 50?

24. Mr. Wilson bought a hat, a coat, and a cloak; he paid for the coat $1\frac{1}{2}$ as much as for the hat, and for the cloak, three times as much as for the coat; and for all he paid 42 dollars; how much did he pay for each?

LESSON XVII.

Analysis of Questions by means of Unity.

1. What number is that to which if 10 be added, the sum will be 16?*

2. James being asked how many marbles he has, replied, if to $\frac{1}{2}$ of what I have you add 8, the sum will be equal to 18: how many had he?

3. What number is that whose half exceeds its third by 4?

4. What number is that, to which if $\frac{1}{2}$ of itself and 4 be added, the sum will be equal to 22?

5. What number is that, which being added to one third of itself, and to 3 times itself, will give a sum equal to 26?

6. What number is that, which being diminished by its half and its third, the remainder will be equal to 4?

---

* Analysis.—Denote the required number by unity. Then by the conditions of the question, unity plus 10 equals 16. But if unity plus 10 equals 16, unity must be equal to 16 minus 10, or 6.

For, *if the same number be subtracted from two equal numbers, the remainders will be equal.*

7. What number is that, to the one fourth of which if 10 be added, the sum will be equal to 20 ?

8. What number is that to which if its one fourth be added, and the sum diminished by 7 will leave 13 for a remainder ?

9. A tailor cuts a piece of cloth 21 yards long into 3 pieces; the second contains $\frac{1}{2}$ of the first minus 5 yards, and the third is equal to $\frac{1}{4}$ of the first plus 5 yards: how many yards are there in each piece ?*

10. One fourth of William's age is equal to one half of John's, and the sum of their ages is 24: what is the age of each ?

11. If one half of Charles' age equals one sixth of John's, and the sum of their ages is 16, what is the age of each ?

12. Divide 15 into two such parts, that one eleventh of the first shall be equal to $\frac{1}{4}$ of the second.

13. A watch and seal are together worth 64 dollars, and the watch is worth 7 times as much as the seal: what is the value of each ?

14. A mother divided 56 pins between Jane and Nancy, so that one fifth of Jane's was equal to one half of Nancy's: how many had each ?

---

* ANALYSIS.—Denote the first part by unity. Then

unity $=$ the first part,

$\frac{1}{2}$ of unity $- 5 =$ the second part, and

$\frac{1}{4}$ of unity $+ 5 =$ the third part; and since

the sum of the parts is equal to 21 yards,

unity $+ \frac{1}{2}$ unity $- 5 + \frac{1}{4}$ of unity $+ 5 = 21$; that is,

$\frac{7}{4}$ of unity $= 21$; and

$\frac{1}{4}$ of unity $= 3$, or unity $= 12$.

Hence, the pieces are 12, 1, and 8 yards.

15. John gave one third of his marbles to William, and then gave half of what he had left and 4 more to Charles, after which he had 8 remaining: how many had he at first?

16. James and John together, have 50 marbles; 3 times James' number is equal to 7 times John's: how many has each?

17. The sum of two numbers is 16, and the greater is 3 times the smaller: what are the numbers?

18. Two partners in trade have made a profit of 99 dollars, and agree to divide it so that the second shall have 4 dollars every time the other has 5: what was the portion of each?

19. A school of 88 scholars has three classes; the second contains $1\frac{1}{2}$ times as many as the first, and the third twice as many as the second: how many scholars in each class?

20. If $\frac{1}{6}$ of John's marbles is equal to $\frac{1}{8}$ of James', and together they have 56, how many has each?

21. A piece of cloth is divided into 3 parts; one piece is 4 yards long, which is one eighth of the length of the other two, but of these two pieces the longer is 3 times the shorter: what is the length of each piece?

22. Two persons, A and B, at a tavern, spend 80 cents, of which twice what A spends is equal to 3 times what B spends: how much is spent by each?

23. The sum of the ages of two persons is 56 years, and twice the age of the elder is 6 times the age of the younger: what is the age of each?

24. A man has 66 fowls, and after selling a part of them found that what he had left was twice the number sold: how many did he sell, and how many had he left?

25. A man after spending a part of his money at a tavern, found that what he had left was one fifth of

what he had spent, and remembering that he had 84 cents at first, wished to know how much he had left?

26. A man sold a horse for 140 dollars, by which he gained $\frac{2}{5}$ of what the horse cost him: what did he give for the horse?

27. John and Charles receives 18 cents for premiums, t school. If three times John's money be subtracted from 3 times what both receive, the remainder will be 24: how much does each receive?

28. A fish weighs 64 pounds. The head weighs 3 times as much as the tail, and the body weighs as much as the head and tail both: what is the weight of each part?

29. If a shadow 15 feet long is cast by a stick 10 feet long, standing vertically, what will be the length of a stick or pole, like placed, which casts a shadow 24 feet long, at the same time of day?

30. If a shadow 10 feet long is cast by a stick 5 feet high, what will be the length of a stick, standing in a like position, which casts a shadow 16 feet in length, at the same time of day?

31. A tailor cut 6 coats from a piece of cloth, after which it measured 24 yards; he then cut 5 pairs of pantaloons, which took $\frac{1}{2}$ as much as the coats, when it was found that one half of the piece was left: how many yards did the piece contain?

32. A farmer has two bins for grain, which together, hold 60 bushels; if he diminishes the greater by $\frac{1}{6}$ of its capacity and increases the less by the same amoun; the two will then hold an equal number of bushels: how many bushels does each hold?

LESSON XVIII.

Promiscuous Questions.

1. A man sold a barrel of flour for \$4, which was $\frac{2}{3}$ of what it cost: how much profit did he make?

2. If $\frac{8}{9}$ of a piece of broadcloth, containing 18 yards, cost 32 dollars, how much was that a yard?

3. A man sold a cow for 18 dollars, which was $\frac{6}{7}$ of what she was worth: what was her true value?

4. A man spends $\frac{1}{5}$ of his monthly income for a hat, and twice as much for a coat, and has \$10 left: how much does he receive a month?

5. A man pays $\frac{1}{4}$ of his daily wages for board, and $\frac{1}{2}$ for his clothes, and at the end of the week has saved \$1$\frac{1}{2}$: what are his wages a day?

6. A pole stands $\frac{1}{2}$ in the air, $\frac{1}{3}$ in the water, and 3 feet in the mud: how long is it?

7. The body of a fish is three times as long as his head, and his tail is 2 feet, which is one fourth the length of the head and body: what is the length of each part, and of the entire fish?

8. A person gave $\frac{1}{6}$ of his money to each of 5 persons, and had 4 cents left: how much had he at first?

9. John gave Charles 4 times as many apples as he gave to William, and to William $\frac{1}{5}$ as many as he had left, which was ten: how many had he at first?

10. What will be the cost of a bag of coffee, if $\frac{2}{5}$ of it cost \$4$\frac{2}{5}$?

11. Mary gave $\frac{3}{4}$ of her money to Jane, and $\frac{1}{6}$ to Eliza, and had 3 cents remaining: how much had she at first?

12. If 3 men can do a piece of work in 8 days, how long will it take 12 men to do the same work?

13. If 5 men can do a piece of work in 7$\frac{1}{5}$ days, how long will it take nine men to do the same work? How long 12 men? 18 men? 36 men? 72 men?

14. A man would give $\frac{1}{3}$, $\frac{1}{4}$, and $\frac{1}{5}$ of the money in his purse to the oldest, the second, and youngest son; now, if he has fifty-two dollars left, how much will each receive?

15. James and John, on comparing their marbles, find that together, they have 112; and that if James gives $\frac{1}{8}$ of his to John, they will have an equal number: how many has each?

16. James, John, and Charles together, have 120 marbles; if James gives $\frac{1}{6}$ of his to John, each will have $\frac{1}{2}$ as many as Charles: how many marbles has each?*

17. James, John, and Charles together, have 300 marbles; if James gives 12 more than $\frac{1}{7}$ of his to John, each will have $\frac{1}{3}$ as many as Charles: how many has each?

18. In a fruit orchard, one third of all the trees are apples, $\frac{1}{4}$ are pears, $\frac{1}{5}$ are cherries, and 26 are plums: how many of each sort, and how many in all?

19. A person being asked the time of day, said, that the time past 12 o'clock M. (that is noon), was $\frac{1}{4}$ the time past the previous midnight: what was the time?

How many equal parts (fourths), from 12, midnight, to the required time? How many of these parts between midnight and noon? What is the value of each part?

20. How will you divide half a water-melon among 3 boys, so that the second shall have twice as much as the first, and the third twice as much as the second: what part of the whole melon will each boy have?

NOTE.—After James has given one sixth of his marbles to John, each will have one fourth of 120, or 30, and Charles will have one half of 120 or 60. Then, what number diminished by $\frac{1}{6}$ leaves 30?

21. If $\frac{2}{3}$ of a barrel of flour will last a family of 5 persons 10 days, how long will a barrel last a family of 25 persons?

22. If $5\frac{2}{3}$ baskets of peaches are worth $\$8\frac{1}{2}$, how many potatoes at 25 cents a bushel, will one basket buy?

23. If 25 bushels of oats will serve 5 horses for 7 days, how many will serve 7 horses for 9 days?

24. A farmer has his sheep in 3 partures. In the first he has $\frac{1}{6}$ of his flock, in the second $\frac{1}{4}$, in the third $\frac{1}{2}$, and 6 over: how many has he in all?

25. A tailor has a piece of cloth, from which he cuts enough for a suit of clothes, and finds that he has 28 yards left, which is just $\frac{1}{2}$ of $\frac{2}{3}$ of his piece: how many yards in the piece?

26. A man starts on foot from Albany, for New York, and at the end of the first day finds that he has 125 miles yet to travel, which was just $\frac{5}{6}$ of the whole distance: what was the whole distance?

27. A grocer bought an equal number of lemons and oranges; he paid 9 cents for every 2 oranges, and 7 cents for every 4 lemons: what must he sell them at a piece to make 100 per cent?

28. A grocer bought a certain number of eggs at the rate of 2 for 5 cents, and an equal number at the rate of 3 for 7 cents, and sold them at 3 cents a piece, by which he made 21 cents: how many eggs did he buy?

29. A man and his wife consumed ten pounds of meat in 3 days. The man alone would have consumed it in 5 days: what part of the meat did the woman consume?

30. If two men can dig 32 bushels of potatoes in 1 day, working 8 hours a day, how long will it take 3 men, working 9 hours a day, to dig 54 bushels?

31. If 5 men can build a fence 20 rods long in 8 days, how long will it require 16 men to build it?

32. What number is that from which if 2 be subtracted $\frac{2}{3}$ of the remainder will be 4? What number is that $\frac{2}{3}$ of which is 4? If 2 be added, what number?

33. After paying away $\frac{1}{3}$ and $\frac{1}{4}$ of my money, I had 10 dollars remaining: how much had I at first?

34. A person after spending $\frac{1}{2}$ of his money, and then $\frac{1}{3}$ of the remainder, had 8 dollars left: how much had he at first?

35. A man at play lost $\frac{1}{3}$ of his money, and the next night lost $\frac{1}{4}$ of the remainder, when he found that he had but $12 left: how much had he at first?

36. John is 20 paces ahead of Charles, but Charles takes 2 steps while John takes 1: how many steps will John make before he is overtaken? How much does Charles gain at each step?

37. If John is 15 steps in advance, and Charles makes 3 steps while John makes 2: how many steps will Charles make before he comes up?

38. John being asked how many marbles he had, said, I have but four, for James first took $\frac{1}{3}$ of all I had, and Charles then took $\frac{1}{3}$ of what was left: how many had he before he lost any?

39. A man bought a harness, a carriage, and a pair of horses. The horses cost 70 dollars more than the carriage, and the carriage 50 dollars more than the harness, which cost 40 dollars: what was the whole cost?

40. A tailor cut a piece of cloth into 3 parts: the first part contained 2 yards more than the second; the second contained 4 yards less than the third; and the third was found to contain 12 yards: how many yards in the piece?

41. A tailor cut a piece of cloth into 3 parts: the

first part contained $\frac{1}{4}$ of the piece and 4 yards over; the second contained $\frac{1}{2}$ of the piece wanting 2 yards; and the third contained $\frac{1}{5}$ of the piece and 6 yards over: how many yards in the whole piece, and how many in each part?

42. A man paid $\frac{3}{5}$ of his year's bill; after which he paid $\frac{1}{3}$ of what was left, and yet owed 12 dollars: how much was the bill?

43. A man spends in a pleasure trip $\frac{3}{5}$ of his money at a hotel, $\frac{1}{20}$ of it in a railroad fare, and 14 dollars in carriage hire and other expenses: how much did he spend in all?

44. Mr. Wilson spends $\frac{3}{5}$ of his money at a tavern, and then goes to a grocery and pays $\frac{2}{3}$ of the remainder for coffee and tea; he counted what he had left, and found 2 dollars: how much had he at first?

45. A young lady had a portion at marriage. She expended $\frac{4}{5}$ of it in furniture, gave 6 dollars to each of two sisters, and had 28 dollars left: how much had she at first?

46. James has half as many marbles as Charles and 10 over; William has half as many as James, and Charles has as many as James and William: how many has each?

47. A tailor wishes to divide a piece of cloth containing 50 yards, into two such parts that one part shall be $5\frac{1}{4}$ times the other: how many yards in each piece?

48. A farmer has a field containing 30 acres of land, and wishes to divide it into 3 such fields that the second shall be double the first, and the third equal to the sum of the other two: how much in each of the new fields?

49. A father buys 70 marbles for John and Henry; and wishes that John should have $2\frac{1}{2}$ times as many as Henry: how must he divide them?

50. A tailor's bill amounts to 96 dollars; 4 articles were charged, vests, pantaloons, and coats; there was 3 times as much charged for pantaloons as for vests, and twice as much for coats as for vests and pantaloons: how much was charged for each?

51. A tailor having a piece of cloth, cut it into two parts, one of which was 2 yards less than one half the piece; he then cut the smaller piece into two equal parts, when he found that each part contained 14 yards: how many yards in the piece?

52. A mason built a wall in 2 days. The first day he built $\frac{1}{3}$ of it and 2 rods over; the second day he built $\frac{1}{2}$ of the remainder and 3 rods more: what was the length of the wall?

53. A horse is worth $4\frac{1}{2}$ times the saddle, and both are worth 110 dollars: what is the value of each?

54. A father divided a farm of 130 acres between 3 sons; the second was to have $1\frac{1}{2}$ times as much as the first, and the third to have $\frac{1}{2}$ as much as the second: what was the share of each?

55. A merchant in settling up his cash account, found that if he had $\frac{1}{3}$ and $\frac{1}{2}$ more, that he would still need 4 dollars to make $70: how much had he?

56. The difference between $\frac{2}{3}$ and $\frac{3}{5}$ of a number is 3 less than $\frac{1}{6}$ of the number: what is the number?

57. The difference between $\frac{5}{6}$ and $\frac{3}{8}$ of a number is 5 greater than $\frac{1}{4}$ of the number: what is the number?

58. A man on foot is 35 miles in advance of a man pursuing him on horseback; the footman travels $3\frac{1}{2}$ miles an hour, and the horseman rides 7: how long before the footman will be overtaken?

59. A dog pursues a fox, which is 10 rods in advance; while the fox runs 3 rods, the dog runs 5: how many rods will the dog run before overtaking the fox?

60. A cistern is filled by a pipe which runs 5 gallons a minute; while the water is discharged by a leak, at the rate of 2 gallons a minute: if the cistern holds 120 gallons, how long will it take to fill it, and how many gallons will have run out?

61. The minute hand of a clock moves 12 times as fast as the hour hand, and moves over one space on the face, in five minutes: how long will it take the minute hand to overtake the hour hand when it is one space behind?

62. A poultry-yard contains geese, ducks, and 15 turkeys; if there were 10 more ducks the number would be equal to that of the geese and turkeys; and the number of the geese is equal to $\frac{5}{6}$ the number of the ducks: how many are there of each sort?

63. A father distributed a sum of money between his three sons, thus: to John he gave $\frac{3}{8}$ of the whole and 9 dollars over; to Reuben he gave 15 dollars; and to William he gave the remainder, which was $\frac{1}{3}$ of the sum that he gave to his other two sons: how much money did he distribute?

64. A piece of cloth containing 86 yards, is cut into two parts: $\frac{2}{9}$ of the whole piece is equal to $\frac{2}{3}$ of the smaller: how many yards must be added to the less piece to make the two pieces equal?

65. A person being asked his age, replied: $\frac{3}{8}$ of my age added to 20 equals my age diminished by 5.

66. A man bought a horse and a colt; $2\frac{1}{2}$ times what he paid for the colt equalled $1\frac{1}{2}$ times 50 dollars, which he paid for the horse: what did he pay for the colt, and how much more for the horse than the colt?

67. A farmer bought a cow and a sheep, and paid 40 dollars for both. He paid for the sheep one half what he paid for the cow, less 8 dollars: what did he pay for each?

68. James has 8 dollars more money than John; $5\frac{1}{2}$ times this difference equals $1\frac{5}{6}$ times James': how much has each?

69. Charles after eating $\frac{1}{4}$ of his chestnuts gave away $\frac{1}{3}$ of what he had left, and then had 16 remaining: how many had he at first?

70. A and B enter into trade together; A puts in 2 dollars every time B puts in 5: A's money remains in 12 months and B's 8; they make a profit of 128 dollars: how should it be divided between them?

71. James went out hunting, and shot one of every 5 squirrels which he saw; had he seen 10 more, and killed in the same proportion, he would have brought home 6: how many squirrels did he see?

72. Two men hired a pasture for $33, and agreed that the pasture of 2 cows should count for 1 horse: one pastured 4 cows and 2 horses for 3 weeks, the other 2 cows and 3 horses for 2 weeks: how much should each pay?

73. If 40 dollars be divided between two persons, so that one shall have 3 dollars every time that the other has 2, how much will each receive?

74. In a school of 44 pupils, there are $1\frac{3}{4}$ times as many girls as boys: how many of each?

75. A man bought a horse, saddle, and bridle: he paid 2 times as much for the saddle as for the bridle, and 11 times as much for the horse as for saddle and bridle both; in all he paid $108: how much did he pay for each?

76. A merchant sold $6\frac{3}{4}$ yards of cloth at 4 dollars a yard, and took his pay in equal quantities of rye and wheat, the former at 50 cents, and the latter at $1 a bushel: how much wheat did he receive?

77. Find the ages of three persons, knowing that the age of the second is equal to twice the age of the first, and the age of the third five times the age of

the first and second, and that the sum of their ages is 90 years.

78. A drover after selling $\frac{3}{5}$ of his flock of sheep, finds that if he had sold 4 less he would have sold just $\frac{1}{2}$ of his flock: how many had he?

79. A boy being asked his age, replied: if to my age you add $\frac{1}{2}$ of it, $\frac{1}{3}$ of it, and 14 years, you will have a sum equal to 3 times my age.

80. A hare pursued by a hound, runs 13 times as far as the distance between them when the pursuit commenced, and the hound runs 28 rods before overtaking the hare: how far was the hare in advance when the pursuit began?

81. A farmer buys a pig, for which he pays 3 dollars, and also a sheep and a cow; the cow cost 3 times as much as the pig and sheep, and the sheep cost 5 times as much as the pig: what was the cost of each?

82. A garrison of 300 men, has provisions for 6 months, at the rate of 16 ounces a day: how much must the allowance be diminished to last 8 months?

83. If a staff 3 feet high, casts a shadow 6 feet in length: how long is a pole which casts a shadow 20 feet long, at the same time of day?

84. If a stick 12 feet long casts a shadow 2 feet long, what is the length of a pole which casts a shadow 9 feet long, at the same time of day?

85. A man's coat cost him $1\frac{1}{2}$ times as much as his pantaloons, his pantaloons cost 6 dollars less than his coat: what was the cost of the pantaloons? What was the cost of the coat?

86. James, John, and Charles together have 150 marbles; Charles has $\frac{1}{4}$ as many as John; if John gives Charles 30 of his, they will all have the same number: how many has each?

it in $2\frac{1}{2}$ hours, and the other in $3\frac{1}{2}$: in what time will they fill it running together?

88. A person hired a man and two boys. To the man he gave six shillings a day, to one boy four, and to the other three; at the end of the time he paid them 104 shillings: how long did they work?

89. A cask of wine leaked out one quarter, after which one third of the remainder was drawn, when the cask was found to contain 30 gallons: how much did the cask hold?

90. A market woman bought a certain number of eggs at 3 for 2 cents, and an equal number at 5 for 4 cents. She paid for both lots 44 cents: how much did her eggs cost her apiece, and how many did she buy?

91. A market woman bought a certain number of eggs at the rate of 4 for 3 cents, and sold them at the rate of 5 for 4 cents, by which she made 4 cents. What did she pay apiece for the eggs? What did she make on each egg sold? How many did she sell to make 4 cents?

92. A market woman bought 36 fowls, of three different sorts, for which she paid 84 shillings. She bought half as many of the first sort as of the second, and three times as many of the third sort as of the first, and paid 1, 2, and 3 shillings apiece for each sort: how many of each did she buy?

93. If James can weed his father's onions in 9 hours, and John in 12 hours, how long will it take both, working together, to weed them?

94. A lady wishes a dress, and does not know whether to buy silk or muslin. The silk costs 9 shillings a yard, and the muslin 3. If she purchases the silk it will cost 72 shillings more than the muslin: how many yards did she need?

he worked he was to receive 5 dimes, and for every day he played he was to pay 2 dimes for his board. When he came to settle he received 87 dimes: how many days did he work?

96. How many small cubes, 1 inch on a side, can be sawed out of a cubic block, 2 feet on a side, allowing no waste in sawing?

97. If A can do $\frac{1}{2}$ of a piece of work in 2 days, and B can do $\frac{1}{4}$ of it in 4 days, how much of it can each do in 1 day, and how long will it take both to do it, working together?

98. A man having a goose, pig, and calf, was asked the value of them. He said that the three were worth 30 shillings, that the goose was worth one third as much as the pig, and that the calf was worth $1\frac{1}{2}$ times as much as the goose and pig together: what was the value of each?

99. James is 10 of his own paces behind John, and in pursuit of him. James steps 3 times while John steps 4 times; but James' steps are twice as long as John's: how many steps must James make to over take him?

100. Two families bought a barrel of flour together, for which they paid $8, and agreed that each child should count half as much as a grown person. In one family, there were 3 grown persons and 3 children; and in the other, 4 grown persons and 10 children; the first family fed from the flour 2 weeks, and the second 3: how much ought each to pay?

www.ingramcontent.com/pod-product-compliance
Lightning Source LLC
LaVergne TN
LVHW011238110826
845150LV00006B/1668

* 9 7 8 1 4 2 5 5 1 3 6 2 7 *